Frontiers in Clinical Drug Research - CNS and Neurological Disorders

(Volume 7)

Edited by

Atta-ur-Rahman, *FRS*

Kings College
University of Cambridge, Cambridge
UK

&

Zareen Amtul

University of Western Ontario
Department of Anatomy & Cell Biology
Schulich School of Medicine & Dentistry
London, ON
Canada

Frontiers in Clinical Drug Research - CNS and Neurological Disorders

Volume # 7

Editor: Atta-ur-Rahman, *FRS* and Zareen Amtul

ISBN (Online): 978-981-14-4752-5

ISBN (Print): 978-981-14-4750-1

ISBN (Paperback): 978-981-14-4751-8

©2020, Bentham Books imprint.

Published by Bentham Science Publishers Pte. Ltd. Singapore. All Rights Reserved.

need for a court order if at any point you breach any terms of this License Agreement. In no event will any delay or failure by Bentham Science Publishers in enforcing your compliance with this License Agreement constitute a waiver of any of its rights.

3. You acknowledge that you have read this License Agreement, and agree to be bound by its terms and conditions. To the extent that any other terms and conditions presented on any website of Bentham Science Publishers conflict with, or are inconsistent with, the terms and conditions set out in this License Agreement, you acknowledge that the terms and conditions set out in this License Agreement shall prevail.

Bentham Science Publishers Pte. Ltd.
80 Robinson Road #02-00
Singapore 068898
Singapore
Email: subscriptions@benthamscience.net

BENTHAM SCIENCE

CONTENTS

PREFACE

The Central Nervous System is undeniably one of the most complex systems of the body. Neuroscientists all over the world are busy exploring this fascinating mystery of the biological system. There has been a tremendous increase in our knowledge about the brain, spinal cord, and peripheral system during the last couple of decades. Yet we still learn something new every day about the multifold obscurities of the nervous system.

The present book is an effort to inform our readers with the new milestones being explored and uncovered every day in the field of neuroscience. The six chapters cover state of the art cutting-edge contributions from eminent leaders in the field. Chapter 1 covers the role of fatty acid amides as new potential therapeutic agents to treat multiple sclerosis; chapter 2 highlights the use of machine learning techniques to detect epileptic seizures; chapter 3 outlines the present and future directions about the essential tremor neurodegeneration in essential tremor; chapter 4 presents the potential therapeutic role of the melatoninergic system to treat epilepsy and comorbid depression; chapter 5 evaluates how the transgenic model of drosophila is being used to model neurodegenerative diseases, and chapter 6 summarizes the importance of genetic basis in stroke as the potential drug target. In brief, each chapter covers a wide-ranging, analytically evaluated body of work such as pathogenesis, therapeutic evidence, targets, and mechanisms of action of therapeutics to treat brain disorders in a very compelling way.

I believe that this book will be of considerable interest for both experienced scientists from the neuroscience community as well as for beginners in the field.

I am grateful for the timely efforts made by the editorial personnel, especially Mr. Mahmood Alam (Director Publications), and Mrs. Salma Sarfaraz (Senior Manager Publications) at Bentham Science Publishers.

Atta-ur-Rahman, *FRS*
Honorary Life Fellow
Kings College
University of Cambridge
Cambridge
UK

&

Zareen Amtul
The University of Windsor
Department of Chemistry and Biochemistry
Windsor, ON
Canada

List of Contributors

Amandine Nachtergael — Unit of Therapeutic Chemistry and Pharmacognosy, University of Mons (UMONS), Belgium

Amarish Kumar Yadav — Cell and Neurobiology Laboratory, Department of Biochemistry, Institute of Science, Banaras Hindu University, Varanasi-221005, India

Anjana Munshi — Department of Human Genetics and Molecular Medicine, Central University of Punjab Bathinda, Punjab, India

Brijesh Singh Chauhan — Cell and Neurobiology Laboratory, Department of Biochemistry, Institute of Science, Banaras Hindu University, Varanasi-221005, India

Dimitrinka Atanasova — Institute of Neurobiology, Bulgarian Academy of Sciences, Sofia1113, Bulgaria
Department of Anatomy, Faculty of Medicine, Trakia University, Stara Zagora6003, Bulgaria

Firas Sabar Miften — University of Thi-Qar, College of Education for Pure Science, Nasiriyah, Iraq

Jonathan H. Green — Faculty of the Humanities, University of the Free State, South Africa
Open Access College, University of Southern Queensland, Australia

Jana Tchekalarova — Institute of Neurobiology, Bulgarian Academy of Sciences, Sofia1113, Bulgaria

Jyotsna Singh — Cell and Neurobiology Laboratory, Department of Biochemistry, Institute of Science, Banaras Hindu University, Varanasi-221005, India

Kanika Vasudeva — Department of Human Genetics and Molecular Medicine, Central University of Punjab Bathinda, Punjab, India

Mario Manto — Department of Neurology, CHU-Charleroi, Chaussée de Bruxelles 140, 6042 Lodelinsart, Belgium

Mohammed Diykh — University of Thi-Qar, College of Education for Pure Science, Nasiriyah, Iraq
School of Agricultural, Computational and Environmental Sciences, University of Southern Queensland, Australia

Nicola S. Orefice — Department of Medicine and Waisman Center, University of Wisconsin, 53705 Madison, USA

Nikolai Lazarov — Institute of Neurobiology, Bulgarian Academy of Sciences, Sofia1113, Bulgaria
Department of Anatomy and Histology, Medical University of Sofia, Sofia1431, Bulgaria

Pierre Duez — Unit of Therapeutic Chemistry and Pharmacognosy, University of Mons (UMONS), Belgium

Rania Aro — Unit of Therapeutic Chemistry and Pharmacognosy, University of Mons (UMONS), Belgium

Rohit Kumar — Cell and Neurobiology Laboratory, Department of Biochemistry, Institute of Science, Banaras Hindu University, Varanasi-221005, India

Roshan Fatima — National Center for Biological Sciences, Bangalore-560097, India

Sangeeta Arya Cell and Neurobiology Laboratory, Department of Biochemistry, Institute of Science, Banaras Hindu University, Varanasi-221005, India

Saripella Srikrishna Cell and Neurobiology Laboratory, Department of Biochemistry, Institute of Science, Banaras Hindu University, Varanasi-221005, India

Shahab Abdulla Open Access College, University of Southern Queensland, Australia

CHAPTER 1

Fatty Acid Amides as a New Potential Therapeutic Agent in Multiple Sclerosis

Nicola S. Orefice[*]

Department of Medicine and Waisman Center, University of Wisconsin, 53705 Madison, USA

Abstract: Multiple sclerosis (MS) is a chronic inflammatory demyelinating disease of the central nervous system (CNS) frequently starting in young adulthood. However, the pathogenesis of the progressive disease phase is still not well-understood, and the inflammation as well as the mechanisms of demyelination and tissue damage is currently being discussed. The available drugs approved in the treatment of different clinical forms of MS prevent the relapses, alleviate the symptoms only partially and slow progression of the disease; however, none of these treatments is capable in stopping the MS clinical course. Moreover, approved MS treatments lead to unpredictable adverse effects associated with a range from mild (such as flu-like symptoms, fatigue, liver transaminase elevation, stomach pain or irritation at an injection site) to serious (such as bradycardia or progressive multifocal leukoencephalopathy). It is time to revise the MS drug development strategy by relying on our endogenous defense mechanisms. Endogenous fatty acid amides (FAAs) are a family of structurally different molecules found in mammalian systems. These compounds include anandamide, oleoylethanolamide and palmitoylethanolamide; research preclinical and clinical reported anti-inflammatory and neuroprotective activity of FAAs making them an alternative therapeutic approach in neurological disorders. In consideration that an endogenous compound able in the control of endogenous defense mechanisms can assume extraordinary importance, this chapter includes a discussion on current approved drugs in MS, and on pharmacological properties of FAAs that may play a promising role in complementing of medication approved for use in MS.

Keywords: Autoimmune disease, Anandamide, Cortical lesions, Endogenous mechanisms, Experimental autoimmune encephalomyelitis, Grey matter, Immune regulatory molecules oleoylethanolamide, Multiple sclerosis, Neuroinflammation, Pain, Palmitoylethanolamide, Preclinical studies, White matter.

[*] **Corresponding author Nicola S Orefice:** Department of Medicine and Waisman Center, University of Wisconsin, Madison, USA; E-mail: nicolaorefice819@gmail.com

Atta-ur-Rahman & Zareen Amtul (Eds.)

INTRODUCTION

Patients living with multiple sclerosis (MS) experience symptoms that negatively affect their quality of life (QOL) when the central nervous system (CNS) disease disrupts nerve signal transmission. It is crucial to supply those who suffer from these symptoms with therapeutic treatments that facilitate healing. Due to the complexity and disease burden of MS, a multidisciplinary management approach that combines pharmacologic and integrative non-pharmacologic therapies is urgently required to provide patients with rapid and effective care.

MEDICAL OVERVIEW OF MULTIPLE SCLEROSIS

Multiple sclerosis (MS) is an inflammatory, autoimmune, demyelinating disease of the central nervous system (CNS) characterized by focal lesions (called plaques) disseminated within of multiple CNS regions. Magnetic resonance imaging (MRI) can accurately detect the demyelinating lesions in white matter (WM), which can be used to provide the clinical diagnosis. The concept that MS is an inflammatory demyelinating disease of WM was established about 50 years ago; recent advances in immunohistochemical staining and MRI sequences have allowed establishing that the focal demyelinating plaques can also damage the grey matter (GM) [1]. Although the MS onset usually occurs in young adults between 20-45 years of age [2], recent clinical studies report MS diagnosis in children and adolescents [3]. MS typically affects young adults, with an initial demyelinating event between 20 years and 40 years of age [3] and has a higher prevalence in women, although it has estimated that more than 10% of persons affected have a history of MS signs or symptoms onset before age 18 [4]. The relapsing-remitting (RR) is the most common clinical form of MS characterized by new attacks (relapses or exacerbations) or a worsening of pre-existing neurologic symptoms associated with a damage of CNS area, followed by periods of partial or complete recovery (remissions). Following this clinical phase, more than half of RR-MS patients switch into a secondary-progressive clinical form of MS (SP-MS), characterized by a worsening of neurological functions (accumulation of disability) independent of acute attack [5]. SP-MS form is characterized by either active phase (with relapses and/or new contrast-enhancing lesion captured by MRI) or not active phase, as well as clinical progression (evidence of disease worsening on an objective measure of change over time, with or without relapses) or without clinical progression. MS patients (approximately 10%) can also experience a disease course characterized by worsening neurologic functions (accumulation of disability) from the onset of symptoms without early relapses or remissions; this clinical condition is defined as primary progressive of MS (PP-MS). Patients affected by progressive MS clinical forms may also exhibit occasional relapses; this subtype of clinical form is classified as progressive-relapsing MS (PR-MS), whereby relapse occurs alongside progression of the disease. Actually, it is still unknown whether MS has a single or multiple causes; nevertheless, factors genetic [6], exposure to virus [7], low exposure to vitamin D [8] may be among the potential causes of MS-related disease activity. Emerging evidence through experimental autoimmune encephamomyelitis (EAE) model, the most commonly used experimental model of MS, has revealed that components of the intestinal microbiome may be involved in autoimmune response, and along this line evidence for a similar cause is beginning to emerge in MS patients [9]. The scientific community is aware that actually non-curative treatment can stop the disease activity as well as the progression of MS. Although, as on December 2017 the Food and Drug Administration (FDA) has approved 15 disease-modifying treatments (DMTs), these medications attenuate the severity of relapse-related effects, and slow but not stop the disability progression. In addition, with the increasing number of medications approved by the FDA, has also increased the risk and the severity of side effects during the treatment. Therefore, the research of new medications capable to rest or slow the MS progression with minimal side effect is becoming increasingly necessary. The current knowledge about the endogenous role of fatty acid amides (FAAs) is taking into consideration the potentiality and effectiveness of this class of neuromodulatory lipids including endogenous cannabinoid N-arachidonoyl ethanolamine (anandamide; AEA), N-palmitoylethanolamine (PEA) and N-oleoylethanolamine (OEA) as therapeutic agents. This chapter highlights the preclinical and clinical outcomes of FAAs making them a promising complementary therapy to the medications currently approved for the treatment of MS-related symptoms.

PATHOGENESIS OF MULTIPLE SCLEROSIS

Actually the exact pathogenesis of MS is still unknown; however the demyelination event is characterized by a lymphocytic (mainly T helper cells) infiltration from periphery to CNS, microglia activation to demyelination and axonal degeneration [10]. Once into the CNS, T- lymphocytes can be reactivated by local professional antigen presenting cells (APCs) like macrophages, microglia and dendritic cells, which are present in human and mouse CNS lesions [11 - 13]. The lymphocytic presence within lesions and bordering areas suggests that inflammatory destruction in MS is driven not only by antigen-specific targeting of myelin, but also by other CNS components like oligodendrocytes, axons, nerve cells and astrocytes [12]. How T-cells become abnormally activated toward CNS antigens remains unclear. In addition to T cells, B cells and their products are involved in the pathogenesis of MS; indeed, it has long been recognized that B cells differentiate into plasma cells to produce antibody molecules closely modeled after the receptors of the precursor B cell. Once released into the blood and lymph node, these antibody molecules bind to the target antigen (foreign substance) and initiate its neutralization or destruction [14, 15]. The presence of these polyclonal antibodies in the cerebrospinal fluid of MS patients is known as oligoclonal bands. The target of these class of antibodies is not yet fully known; however, genetic factors can influence MS pathogenesis susceptibility. Studies of families and twin have shown a 40-fold increased susceptibility among first-degree relatives of MS patients suggesting a genetic basis [16]. Recent studies performed in children and adolescents with MS were focused on the issue of infectious etiology; among the pathogens possibly involved are human herpes virus type 6, Epstein Barr virus, and mycoplasma pneumonia [17].

WHITE MATTER PLAQUE

The WM plaques are detected in predilection sites notably around the ventricles [18]; others predilection sites include the optic nerve and subpial spinal cord [19]. Histologic inspection has also reported WM plaques show poorly defined borders [20]. The WM plaques are defined like chronic and acute active. Chronic plaques are frequently observed than active plaques in MS patients with a progressive phase. They are characterized by a less mononuclear cells, almost complete demyelination and severe astrogliosis [21]. While the active plaques are characterized by ongoing destruction of myelin and are heavily infiltrated by macrophages and microglial cells [22]. In addition, MRI data suggest that acute active plaques represent the pathologic substrate of new clinical attacks [23]. Neuropathological studies have reported that oligodendrocytes cells are preferentially destroyed in early acute plaque [24]; however, oligodendroglial injury is very variable with numerous oligodendrocytes present into the plaque often displaying signs of concurrent early remyelination [24]. On the basis of these specific neuropathological and pathological findings, Lucchinetti and colleagues have classified the WM plaques into four immunopatterns, suggesting that the targets of injury and mechanisms of demyelination in MS may differ between patients [25]. Although MS is considered an inflammatory demyelinating disease of the CNS, axonal injury and loss can also occur in the acute plaque and be associated with the development of permanent disability in MS patients [26, 27]. This suggests that progressive axonal loss may induce transition from RR-MS to SP-MS.

GREY MATTER PLAQUE

Although MS has historically been considered a disease primarily affecting the WM of CNS, this concept has recently been revisited in light of a body of evidence (hystopathological and neuroimaging) establishing that the inflammatory demyelinating lesions can also damage the cortical grey matter (CGM). Three cortical lesion types have been described based on their morphology and location within the cortex: subpial, intra-cortical and leukocortical. Subpial lesion extends from pial surface to cortical layer three or four, or to the entire width of the cortex, and may involve several gyri. Intra-cortical lesions are small, perivascular demyelinated lesions confined within the cortex with the sparing of both superficial cortex and adjacent WM. Leukocortical lesions involve both gray and white matter at the gray matter-white matter junction. Cortical lesions (CLs) are captured using conventional MRI sequences; the introduction of recent imaging protocols using double inversion recovery (DIR) which selectively suppress the signals from cerebrospinal fluid (CSF) and WM, has improved CLs detection in MS patients [28]. The use of these sequences allow to detect the CLs not only in patients with a progressive phenotypes of MS, but also in those with RR clinical form, even at clinical onset [1]. Recent longitudinal studies have suggested a direct impact of CLs on physical and cognitive long-term disability in all MS subsets [29]. Although CLs are characterized by substantial loss of oligodendrocytes and axons, they differ markedly from WM lesions in terms of the degree and type of inflammation. Intracortical lesions typically have a very low degree of inflammation [30 - 32]; perivascular infiltrates are rarely found in MS cortex, and the density of infiltrating lymphocytes in pure CLs is similar to the density of infiltrating lymphocytes in normal appearing GM. Therefore, in CLs the demyelination processes may not be solely immune-mediated.

DISEASE-MODIFYING THERAPIES IN MULTIPLE SCLEROSIS

Current-First-Line Treatment

Beginning with the 1980s, copolymers and interferon were introduced in the treatment of MS; they reduced the rate of relapse with a modest effect on disability progression [33, 34]. These molecules opened the age of disease-modifying therapies (DMTs). The first turning point was in 1996 with the interferon-beta-1b (IFN-β1b) (BETASERON®), the first DMT approved for the treatment of RR-MS clinical form. IFN-β1b is a purified, sterile, lyophilized protein product produced by recombinant DNA techniques. Although, the mechanism of action of IFN-β1b is still unknown, the IFN-β1b receptor binding induces the expression of interferon-induced proteins that are responsible for the pleiotropic bioactivities of the drug. Immunomodulatory effects of IFN-β1b include the enhancement of suppressor T cell activity, reduction of proinflammatory cytokine production, down regulation of antigen presentation, and inhibition of lymphocyte trafficking into the central nervous system. In the following years, interferon-beta-1a (IFN-β1a) (Rebif®) and glatiramer acetat (GA) (Copaxone®) were approved and introduced among the DMTs. IFNs β1a appears to directly increase expression and concentration of anti-inflammatory agents while downregulating the expression of proinflammatory cytokines [35]. IFNs treatment may reduce the trafficking of inflammatory cells across the blood brain barrier (BBB) and increase nerve growth factor (NGF) production, leading to a potential increase in neuronal survival and repair. The mechanistic effects of both IFNs manifest clinically as reduced MRI lesion activity, reduced brain atrophy, increased time to reach clinically definite MS after the onset of neurological symptoms, decreased relapse rate and reduced risk of sustained disability progression. The IFN-β1a and β1a are both formulations for subcutaneous (SC) administration.

GA is a copolymer of four amino acids existing in the myelin basic protein (MBP) administered by SC three times per week. This medication seems to block myelin-damaging T-cells through mechanisms that are still unknown. In addition, a neuroprotective effect possibly mediated by neurotrophic factors such as BDNF, has been proposed in the light of evidence from animal models [36] and in MS patients [37].

According to current guidelines, IFNs and GA are indicated as first-line drugs for the treatment of RR-MS. Teriflunomide (Aubagio®) and dimethyl-fumarate (Tecfidera®) have been approved as for first-line therapy for RR-MS recently. Teriflunomide, as oral drug at two doses of 7 or 14 mg a once-daily in the US and at 14 mg in Europe a once-daily, inhibits proliferating lymphocytes by blocking dihydroorotate-dehydrogenase (DHODH) a mitochondrial enzyme expressed at high levels in proliferating lymphocytes [38]. Additionally, teriflunomide seems to have similar effects on the synthesis of nuclear factor kappa light chain enhancer of activated B cells (NF-κB) [39].

Dimethylfumarate (DMF) is administrated as oral drug (120 mg) twice daily. At the beginning of the second week of treatment, the dose should be increased to the maintenance dose of 240 mg twice a day orally. Although, the exact mechanism of action has been not completely elucidated, DMF may activate a pathway involved in the cellular response to oxidative stress, which is induced by inflammation [40]. Ocrelizumab (Ocrevus®) approved in 2017, is a therapeutic monoclonal antibody that represents a different scientific approach in the treatment of RR clinical form and progressive or worsening MS [41]. It targets a type of immune cell called CD20-positive B cell that plays a key role in the disease [41]. Ocrelizumab is administered once every six months by an intravenous (IV) infusion. Finally, on March 2019 the US Food and Drug Administration (FDA) and European Medicines Agency (EMA) has approved Siponimod (Mayzent®) for the treatment of adults with relapsing forms of MS, including SP-MS with active disease and the RR-MS clinical form [42]. Siponimod is a sphingosine 1-phosphate (S1P) receptor modulator that reduces the migration of lymphocytes into the CNS binding with high affinity to S1P receptors 1 and 5, which are present in CNS cells. It blocks the lymphocytes' capacity to move out from lymph nodes, reducing the lymphocyte count in the peripheral blood and forcing them to move into the CNS. Siponimod is available as round biconvex tablets in 0.25mg and 2mg strengths.

Current-Second-Line Treatment

Due to the lack of a standardized definition of treatment non-response in MS, it is often difficult when to switch from first to second line treatment. Given that relapse activity is a key clinical parameter, a switch in therapy may be required at the earliest sign of relapse activity. However, the current DMTs are unable to fully suppress relapse activity; thus, the only relapse may not be sufficient criteria to switch from first to second line treatment. Natalizumab (Tysabri®) has been the first monoclonal antibody approved for patients with active RR-MS [43, 44]. Natalizumab is designed to block a part of the inflammatory pathway in MS; indeed, the main action is to prevent lymphocytes from crossing BBB blocking adhesion molecules, and administrated in i.v. perfusion every 4 weeks [45]. In 2011 year, EMA approved Fingolimod (Gylenia®), the first once-daily oral drug, to treat highly active in RR-MS [46]. It is a selective immunosuppressant; in particular is a sphingosine1-phosphate receptor modulator that prevents the egress of autoreactive lymphocytes to leave lymphnodes, leading to a reduction of their infiltration into CNS [47]. Alemtuzumab (Lemtrada®) was approved in the EU for adult RR-MS patients with active disease defined by clinical or imaging features in 2013 [48]. Administered by intravenous infusion for five consecutive days initially and for three consecutive days one year later, represents the second monoclonal antibody after Natalizumab. Its action is directed against CD52, a cell-surface protein highly expressed on T and B lymphocytes. The binding of alemtuzumab to CD52 results in the depletion of T and B lymphocytes from the circulation through antibody-dependent cell-mediated cytolysis, complement-dependent cytolysis and induction of apoptosis [49, 50]. (Table **1**) summarizes main side effects.

Table 1. The most common side effects of MS DMTs approved in the US and Europe.

Treatment	MS Clinical Form	Formulation	Adverse effects
Interferon-beta-1b (IFN-β1b)	Relapsing -Remitting	Subcutaneous	**Systemic Reactions**
			flu-like symptoms, leukopenia,
			increased transaminase,increase of spasticity
			Injection site reactions
			erythema, pain, cutaneous reaction

(Table 1) cont.....

			Systemic Reactions
Interferon-beta-1a (IFN-β1a)	Relapsing-Remitting	Intramuscular	elevation of liver enzyme,
			flu-like symptoms, fever, pain, muscle aches
			Injection site reactions
			erythema, swelling, pain
Glatiramer acetate	Relapsing-Remitting	Subcutaneous	**Systemic Reactions**
			Flushing, rash, shortness of breath, chest pain
			Injection site reactions
			redness, warmth, swelling, itching
Teriflunomide	Relapsing -Remitting	Oral	**Systemic Reactions**
			Decreased platelet count, decreased neutrophils,
			increased gamma-glutamyltransferase,
			hyperkalemia, lymphocytopenia
Dimethyl fumarate	Relapsing -Remitting	Oral	**Systemic Reactions**
			Increased transaminase, increased serum aspartate,
			abdominal pain, flushing, skin rash
Ocrelizumab	Primary-Progressive	Intravenous	**Systemic Reactions**
			Progressive Multifocal Leukoencephalopathy,
			hepatitis B virus reactivation,
			weakened immune system
Natalizumab	Relapsing -Remitting	Intravenous	**Systemic Reactions:**
			Progressive Multifocal Leukoencephalopathy,
			hepatotoxicity, herpesviral encephalitis
Fingolimod	Relapsing -Remitting	Oral	**Systemic Reactions**
			Slow heart rate (bradycardia or bradyarrhythmia),
			macular edema, transaminase elevation

(Table 1) cont.....

			Systemic Reactions
Alemtuzumab	Relapsing -Remitting	Intravenous	Decreased white and red blood cells, bronchitis
			thyroid gland, herpes reactivation, *Listeria* meningitis
Siponimod	Secondary-Progressive	Oral	Systemic Reactions
			Increased transaminase, head pain, high blood pressure
			slow heartbeat
Delta-9-tetrahydrocannabinol and cannabidiol	Moderate to severe spasticity due to multiple sclerosis	Oromucosal spray	Adverse events
			Disorientation, euphoric mood,
			dissociation

The first oromucosal spray formulation containing Delta-9-tetrahydrocannabinol (Δ9-THC)/cannabidiol (CBD) at an approximately 1:1 fixed ratio derived from cloned Cannabis sativa L. plants. Sativex® was approved in the United Kingdom in 2010 as second-line therapy for adult patients with moderate-to-severe MS-related spasticity that is resistant to first-line anti-spasticity medications. The main active substance, THC, acts as a partial agonist at human cannabinoid receptors (CB1 and CB2), and thus may modulate the effects of excitatory (glutamate) and inhibitory (gamma-aminobutyric acid) neurotransmitters. An important issue to consider is the lack of clarity about the effect of Sativex® use on cognition that may be attributable to the considerable heterogeneity among studies in terms of cannabis composition. Research indicates that Δ9-THC administration impairs cognition, particularly memory and emotional processing. Limited evidence suggests that CBD administration might improve cognition in cannabis users but not in individuals with neuropsychiatric disorders. In addition, studies indicate that some acute Δ9-THC-induced cognitive impairments may be prevented if Δ9-THC is administered in combination or following CBD treatment. Independent of these potential explanations, the lack of a clear association between cannabis use and impairments in cognition may also due to the considerable heterogeneity in recreational cannabis that participants in these studies may have been exposed, as well as the differing effects on cognition of the various chemicals found in the extract of the cannabis plant. The extract of Cannabis sativa has over 60 different cannabinoids [51], with Δ9-THC and CBD being the most prominent among them. However, while Δ9-THC is thought to be responsible for most of its psychotropic effects [52], CBD is under investigation for its potential antipsychotic effects, in opposition to the pro-psychotic effects of Δ9-THC. Some research suggests that CBD can counteract the negative effects of Δ9-THC, as investigated in both humans and animal models at a behavioral and neurochemical level [53 - 55]. This issue is of crucial importance considering that case-control studies suggest that the risk of development and relapse of psychosis in cannabis users depends on both frequency of use and cannabis potency, with the risk being the highest in individuals exposed on a daily basis to cannabis with a high $\Delta 9$-THC concentration, and unchanged among users of cannabis with a lower $\Delta 9$-THC concentration and a more balanced $\Delta 9$-THC:CBD ratio.

ENDOGENOUS FATTY ACID AMIDES

Although the immune modulating drugs continue to be the main therapy in MS, the endogenous fatty acid amides (FAAs) are being widely considered as new possible therapeutics agents in MS. FAAs represent a class of neuromodulatory lipids that includes the endocannabinoid anandamide, the "endocannabinoid-like" N-oleolylethanolamine (OEA) and N-palmitoylethanolamine (PEA). FAAs are rapidly hydrolyzed *in vivo* [56] by fatty acid amide hydrolase (FAAH) enzyme. Mice with a targeted disruption in the FAAH gene (FAAH(–/–) mice) [56] or those treated with FAAH inhibitors [56] are severely impaired in their ability to degrade FAAs showing hypersensitivity to the pharmacological effects of these lipids. Blockade of FAAH activity also leads to highly elevated endogenous levels of FAAs in the nervous system [56] and peripheral tissues [57] that correlate with analgesic, anxiolytic, and anti-inflammatory phenotypes [58, 59]. The two human FAAH enzymes, which share 20% sequence identity and are referred to hereafter as FAAH-1 and FAAH-2, hydrolyzed primary fatty acid amide substrates (*e.g.* oleamide) at equivalent rates, whereas –FAAH-1 exhibited much greater activity with N-acyl ethanolamines (*e.g.* anandamide). Although, the FAAH activity in mammals has been primarily attributed to a single integral membrane enzyme of the amidase signature (AS) family, Ueda and colleagues [60] identified a distinct N-acylethanolamines (NAE) hydrolase enriched in immune cells that resides in the lysosome and exhibits an acidic pH optimum. This lysosomal "acid" amidase is not an AS enzyme, but rather related to acid ceramidases. The contribution that acid amidase makes to fatty acid amide catabolism *in vivo* remains unknown, although the distinct inhibitor sensitivity profiles of this enzyme and FAAH-1 should allow straightforward pharmacological separation of their respective roles in living systems.

CLASS OF LIPID TRANSMITTERS

Anandamide is known to modulate several biological and behavior processes [61, 62], including body temperature [63], locomotion [64] and pain perception [65]. Although, the majority of biological effects of anandamide result through its high affinity to CB1 cannabinoid receptors, a low affinity has been reported with transient receptor potential vanilloid 1 (TRPV1) [66], peroxisome proliferator-activated receptor alpha and gamma [67] and CB2 cannabinoid receptors [68]. AEA signaling is terminated through catabolism by the endoplasmic reticulum-localized FAAH enzyme [69]. Prior to reaching FAAH, lipophilic AEA requires transport through the aqueous cytosol. To date, several intracellular AEA binding proteins have been identified such as fatty acid binding proteins (FABPs), Hsp70, and most recently FAAH-like anandamide transporter (FLAT) [69]. OEA is a monounsaturated analogue, and functional antagonist of anandamide which, acting through mechanisms independent of CB1 receptors, might be involved in the regulation of different pathophysiological aspects of appetite and metabolism regulation. Indeed, several studies reveal that trough a high affinity with the PPAR-α [70], OEA controls the expression of several genes involved in fat absorption and fatty acid metabolism, to activate hypothalamic oxytocinergic neurons and to inhibit further eating [71]. The hydrolytic cleavage of OEA is catalyzed by FAAH enzyme [72]. PEA has attracted much attention because it exerts a local anti-injury function through a down-modulation of mast cells and protects neurons from excitotoxicity through several mechanisms. In particular, PEA through a PPARα-dependent mechanism, and G protein-coupled receptor 55 (GPR55) [73] exerts anti-inflammatory, analgesic, and neuroprotective actions [74, 75]. PPAR-α actually seems to be the main molecular target involved in the anti(neuro)inflammatory effects of PEA. When activated by a ligand, PPAR-α forms a heterodimer with 9-cis-retinoic acid receptor (RXR) able to interact with specific DNA sequences in the promoter regions of selective genes, thus leading to complex anti-inflammatory responses. The PEA synthesis occurs from phospholipids through the sequential actions of N-acyltransferase and N-acylphosphatidylethanolamine-preferring phospholipase D (NAPE-PLD), and its actions are terminated by its hydrolysis by FAAH and N-acylethanolamin--hydrolysing acid amidase (NAAA) [76]. Probably due to the fact that PEA is an endogenous modulator as well as a compound in food, such as eggs and milk, no serious side effects have been reported, nor have drug-drug interactions. This has allowed making it as an ideal candidate as adjuvant therapy in various clinical trials in neurological disorders [77], and neuropathic pain [78, 79].

ENDOGENOUS FATTY ACID AMIDES IN MULTIPLE SCLEROSIS

Preclinical and Clinical Evidence

Anandamide

How MS or experimental autoimmune encephalomyelitis (EAE) animal model perturb the FAAs concentration is still under discussion. However, the use of selective inhibitors of the cellular uptake of AEA, such as OMDM1 and OMDM2, showed an increase of AEA concentration into spinal cord of Theiler's murine encephalomyelitis virus-induced demyelinating disease (TMEV-IDD) [80], a viral model of MS where the persistent infection leads in susceptible strain of mice to a T-cell mediated acute disease resembling encephalomyelitis followed by late chronic demyelinating disease [81]. Several studies reported that macrophages/microglia infected with TMEV increased their production of AEA [82] and the selective use of AEA inhibitor (UCM-707) showed a decrease of severity of symptoms in TMEV-IDD model with a decrease also of microglia activation and immune cellular infiltration [83]. This bioactive lipid has been also observed to inhibit human Th-1 and Th-17 cells [84] and reduce interleukin IL-23 and IL-12 release by microglial cells by a mechanism that engages endogenous production of the protective cytokine IL-10 [85, 86]. Increased levels of AEA were also detected in the brains of EAE mice during the acute phase of the disease, possibly accounting for its anti-excitotoxic action in this disorder [87]. On the other hand, because AEA can bind to the TRPV1 ion channel [86], which activation may negatively affect neuronal survival? In clinical studies, in line with the data observed in preclinical experiment, a significant alteration of AEA concentration was reported in MS patients. In inflammatory lesions of patients with active MS and in lesions of patients with silent MS, a higher concentration of AEA found in comparison to healthy controls [87]. In keeping with these results, Centonze *et al*, have highlighted the importance of AEA concentrations increased into CSF and in peripheral lymphocytes of RR-MS patients, and in active lesions than in quiescent lesions of MS patients [85].

Oleoylethanolamide

Several authors have reported that OEA is a potent anti-inflammatory and antioxidant compound that exerts neuroprotective effects through PPARα [88, 89]. Paying special attention on neuroinflammation topic in MS, some authors have highlighted that systemic administration of OEA is able to cross the BBB reaching the brain rapidly exerting an antiinflammatory effect [78, 90]. The antiinflammatory action of OEA is related to its PPARα agonist within the brain; since activation of this nuclear receptor has been shown to induce changes in inflammatory-related genes by repressing the nuclear factors NF-κB and AP-1 [91, 92]. Although no clinical trial have not been perform to assess the OEA proprieties in MS, levels of OEA were significantly elevated in SPMS plasma and in patients with MRI gadolinium-enhancing (Gd+) lesions [93].

Palmitoylethanolamide

In chronic models of MS such as chronic relapsing experimental autoimmune encephalomyelitis (CREAE), myelin oligodendrocyte glycoprotein-induced experimental autoimmune encephalomyelitis (MOG-EAE), and TMEV-IDD have been assessed the therapeutic effects of PEA. In addition, it was observed that the exogenous administration of PEA leaded to an ameliorated spasticity or motor deficits in mice with CREAE or TMEV-IDD [94], and reduced the severity of neurobehavioral scores in mice with MOG-induced EAE [84]. Furthermore, the expression of inflammatory cytokines was greatly reduced by PEA both in TMEV-IDD and MOG-induced EAE mice [95], and these effects were accompanied by decreased of demyelination and axonal damage [96]. These data have drawn conclusions that levels of exogenous PEA might be helpful to compensate or amplify the endogenous defense mechanism deployed by the T-cells or tissues to counteract neurodegenerative and neuro-inflammatory processes. Given its poor water solubility, large particle size in the native state, and possibly short-lived action, PEA might have limitations in terms of solubility and bioavailability. In fact, PEA is almost insoluble in water, while its solubility in most other aqueous solvents is very poor with a partition coefficient (log P) estimated to be > 5 [97]. It has recently been demonstrated that the administration of a new formulation including PEA, and the antioxidant compound luteolin (Lut) [98, 99] subjected to an ultramicronization process (co-ultraPEALut) when compared to either molecule alone or in simple combination, exerts superior anti-inflammatory action improving neurological outcome in experimental models of MS. Based on these intriguing findings, several pre-clinical studies have been addressed to establish the effects of co-ultraPEALut on expression of different myelin proteins in oligodendrocyte progenitor cells suggesting. The new PEA-formulation might stimulate the maturation of undifferentiated oligodendrocyte precursor cells (OPCs) *in vitro* [100] and promote remyelination in mouse models of MS [99]. PEA-containing products (Normast®, Glialia®, Nevamast®, Adolene®, Visimast®, Mastocol®, and Pelvilen®) are already licensed for use in humans (generally 1,200 mg/day) as a nutraceutical, food supplement, or food for medical purposes depending on the country. To the best of our knowledge, current clinical studies of PEA are mostly related to pain or peripheral inflammatory-related conditions, while there are very few studies aimed at evaluating the possible beneficial effects of PEA on CNS-related pathologies in human beings. Initially it was observed that endogenous levels of PEA increased in the plasma of SP-MS [86]; another study report endogenous levels of PEA increased in the plasma of SP and RR-MS clinical form [106]. Significantly reduced levels was found in the cerebrospinal fluid (CSF) of patients with MS compared to control, with lower levels detected in the secondary progressive subtype [93]. A clinical study reported that PEA is a safe drug with a well-known toxicological profile and it can be can as orally administered to reduce the cutaneous adverse effects of interferon-1a, and circulating proinflammatory cytokines in RR-MS [78]. The relevance of results from models for MS pathogenesis has been validated [101 - 104] using also a formulation of micronized and ultramicronized N-palmitoylethanolamine [105] obtaining particles with a defined size profile completely different and statistically lower (6–10 μm at most) in comparison to naïve PEA (in the 100–700 μm range). The new PEA formulation has also allowed to enhance the dissolution rate of drug and reduce its variability of absorption when orally administered.

CONCLUSION

It has been more 25 years, and the management of MS has become a neurology success. Advances in understanding of the disease mechanisms and the dynamic nature of the disease have brought the DMTs to market in many countries. However, treatment is still hampered by adverse effects and by limited evidence of efficacy in more advanced 'progressive' MS. Several patients do not receive DMTs for years after diagnosis, or are that medications can no longer help as their disability is worsening. In addition, the cost of DMTs is a significant issue and have to be considered, especially in low-income and middle-income countries. Over the past ten years, a remarkable number of advances made in our understanding of the molecular and physiological functions of FAAs. In particular, preclinical either *in vitro* or *in vivo* data strongly suggest that PEA, especially its ultramicronized formulation, exerts quite robust therapeutic effects. Thus, this chapter invites to look at this non-endocannabinoid lipid mediator like new and emerging therapy opportunities to manage the side effects of disease-modifying therapies in MS; in other words, a new therapeutic approach able to maintain a balance of immune homeostasis.

CONSENT FOR PUBLICATION

Not applicable.

CONFLICT OF INTEREST

The author confirm that this chapter content has no conflict of interest

ACKNOWLEDGEMENT

Declared none.

REFERENCES

[1]   Calabrese M, Favaretto A, Martini V, Gallo P. Grey matter lesions in MS: from histology to clinical implications. Prion 2013; 7(1): 20-7.
       [http://dx.doi.org/10.4161/pri.22580] [PMID: 23093801]

[2]   Filippi M, Bar-Or A, Piehl F, *et al.* Multiple sclerosis. Nat Rev Dis Primers 2018; 4(1): 43.
       [http://dx.doi.org/10.1038/s41572-018-0041-4] [PMID: 30410033]

[3]   Waldman A, Ghezzi A, Bar-Or A, Mikaeloff Y, Tardieu M, Banwell B. Multiple sclerosis in children:

an update on clinical diagnosis, therapeutic strategies, and research. Lancet Neurol 2014; 13(9): 936-48.
[http://dx.doi.org/10.1016/S1474-4422(14)70093-6] [PMID: 25142460]

[4]   Charvet L, Cersosimo B, Schwarz C, Belman A, Krupp LB. Behavioral Symptoms in Pediatric Multiple Sclerosis: Relation to Fatigue and Cognitive Impairment. J Child Neurol 2016; 31(8): 1062-7.
[http://dx.doi.org/10.1177/0883073816636227] [PMID: 26961266]

[5]   Goldenberg MM. Multiple sclerosis review. P&T 2012; 37(3): 175-84.
[PMID: 22605909]

[6]   Oksenberg JR, Begovich AB, Erlich HA, Steinman L. Genetic factors in multiple sclerosis. JAMA 1993; 270(19): 2362-9.
[http://dx.doi.org/10.1001/jama.1993.03510190118037] [PMID: 8230601]

[7]   Djelilovic-Vranic J, Alajbegovic A. Role of early viral infections in development of multiple sclerosis. Med Arch 2012; pp. 37-40.

[8]   Sintzel MB, Rametta M, Reder AT. Vitamin D and Multiple Sclerosis: A Comprehensive Review. Neurol Ther 2018; 7(1): 59-85.
[http://dx.doi.org/10.1007/s40120-017-0086-4] [PMID: 29243029]

[9]   Li XL, Zhang B, Sun MJ, *et al.* Mechanism of gut microbiota and Axl/SOCS3 in experimental autoimmune encephalomyelitis. Biosci Rep 2019;. pp. 39(7)

[10]   Luo C, Jian C, Liao Y, *et al.* The role of microglia in multiple sclerosis. Neuropsychiatr Dis Treat 2017; 13: 1661-7.
[http://dx.doi.org/10.2147/NDT.S140634] [PMID: 28721047]

[11]   Greter M, Heppner FL, Lemos MP, *et al.* Dendritic cells permit immune invasion of the CNS in an animal model of multiple sclerosis. Nat Med 2005; 11(3): 328-34.
[http://dx.doi.org/10.1038/nm1197] [PMID: 15735653]

[12]   Frohman EM, Racke MK, Raine CS. Multiple sclerosis--the plaque and its pathogenesis. N Engl J Med 2006; 354(9): 942-55.
[http://dx.doi.org/10.1056/NEJMra052130] [PMID: 16510748]

[13]   Pierson ER, Stromnes IM, Goverman JM. B cells promote induction of experimental autoimmune encephalomyelitis by facilitating reactivation of T cells in the central nervous system. J Immunol 2014; 192(3): 929-39.
[http://dx.doi.org/10.4049/jimmunol.1302171] [PMID: 24367024]

[14]   Hofmann K, Clauder AK, Manz RA, Targeting B. Targeting B Cells and Plasma Cells in Autoimmune Diseases. Front Immunol 2018; 9: 835.
[http://dx.doi.org/10.3389/fimmu.2018.00835] [PMID: 29740441]

[15]   Pröbstel AK, Sanderson NS, Derfuss T. B Cells and Autoantibodies in Multiple Sclerosis. Int J Mol Sci 2015; 16(7): 16576-92.
[http://dx.doi.org/10.3390/ijms160716576] [PMID: 26197319]

[16]   Milo R, Kahana E. Multiple sclerosis: geoepidemiology, genetics and the environment. Autoimmun 2010; pp. 387-A394.

[17]   Fujinami RS, von Herrath MG, Christen U, Whitton JL. Molecular mimicry, bystander activation, or viral persistence: infections and autoimmune disease. Clin Microbiol Rev 2006; 19(1): 80-94.
[http://dx.doi.org/10.1128/CMR.19.1.80-94.2006] [PMID: 16418524]

[18]   Popescu BF, Pirko I, Lucchinetti CF. Pathology of multiple sclerosis: where do we stand? Continuum (Minneap Minn) 2013; 19(4 Multiple Sclerosis): 901-21.
[http://dx.doi.org/10.1212/01.CON.0000433291.23091.65] [PMID: 23917093]

[19]   Bradl M, Reindl M, Lassmann H. Mechanisms for lesion localization in neuromyelitis optica spectrum disorders. Curr Opin Neurol 2018; 31(3): 325-33.

[http://dx.doi.org/10.1097/WCO.0000000000000551] [PMID: 29465432]

[20]   Kipp M, van der Valk P, Amor S. Pathology of multiple sclerosis. CNS Neurol Disord Drug Targets 2012; 11(5): 506-17.
[http://dx.doi.org/10.2174/187152712801661248] [PMID: 22583433]

[21]   Ponath G, Park C, Pitt D. The Role of Astrocytes in Multiple Sclerosis. Front Immunol 2018; 9: 217.
[http://dx.doi.org/10.3389/fimmu.2018.00217] [PMID: 29515568]

[22]   Wu GF, Alvarez E. The immunopathophysiology of multiple sclerosis. Neurol Clin 2011; 29(2): 257-78.
[http://dx.doi.org/10.1016/j.ncl.2010.12.009] [PMID: 21439440]

[23]   Frischer JM, Weigand SD, Guo Y, *et al.* Clinical and pathological insights into the dynamic nature of the white matter multiple sclerosis plaque. Ann Neurol 2015; 78(5): 710-21.
[http://dx.doi.org/10.1002/ana.24497] [PMID: 26239536]

[24]   Patrikios P, Stadelmann C, Kutzelnigg A, *et al.* Remyelination is extensive in a subset of multiple sclerosis patients. Brain 2006; 129(Pt 12): 3165-72.
[http://dx.doi.org/10.1093/brain/awl217] [PMID: 16921173]

[25]   Metz I, Weigand SD, Popescu BF, *et al.* Pathologic heterogeneity persists in early active multiple sclerosis lesions. Ann Neurol 2014; 75(5): 728-38.
[http://dx.doi.org/10.1002/ana.24163] [PMID: 24771535]

[26]   Trapp BD, Peterson J, Ransohoff RM, Rudick R, Mörk S, Bö L. Axonal transection in the lesions of multiple sclerosis. N Engl J Med 1998; 338(5): 278-85.
[http://dx.doi.org/10.1056/NEJM199801293380502] [PMID: 9445407]

[27]   Bjartmar C, Kidd G, Mörk S, Rudick R, Trapp BD. Neurological disability correlates with spinal cord axonal loss and reduced N-acetyl aspartate in chronic multiple sclerosis patients. Ann Neurol 2000; 48(6): 893-901.
[http://dx.doi.org/10.1002/1531-8249(200012)48:6<893::AID-ANA10>3.0.CO;2-B]   [PMID: 11117546]

[28]   Gabr RE, Sun X, Pednekar AS, Narayana PA. Automated patient-specific optimization of three-dimensional double-inversion recovery magnetic resonance imaging. Magn Reson Med 2016; 75(2): 585-93.
[http://dx.doi.org/10.1002/mrm.25616] [PMID: 25761973]

[29]   Messina S, Patti F. Gray matters in multiple sclerosis: cognitive impairment and structural MRI. Mult Scler Int 2014; 2014609694
[http://dx.doi.org/10.1155/2014/609694] [PMID: 24587905]

[30]   Peterson JW, Bö L, Mörk S, Chang A, Trapp BD. Transected neurites, apoptotic neurons, and reduced inflammation in cortical multiple sclerosis lesions. Ann Neurol 2001; 50(3): 389-400.
[http://dx.doi.org/10.1002/ana.1123] [PMID: 11558796]

[31]   Bø L, Vedeler CA, Nyland H, Trapp BD, Mørk SJ. Intracortical multiple sclerosis lesions are not associated with increased lymphocyte infiltration. Mult Scler 2003; 9(4): 323-31.
[http://dx.doi.org/10.1191/1352458503ms917oa] [PMID: 12926836]

[32]   Kutzelnigg A, Lucchinetti CF, Stadelmann C, *et al.* Cortical demyelination and diffuse white matter injury in multiple sclerosis. Brain 2005; 128(Pt 11): 2705-12.
[http://dx.doi.org/10.1093/brain/awh641] [PMID: 16230320]

[33]   Arnon R, Aharoni R. Mechanism of action of glatiramer acetate in multiple sclerosis and its potential for the development of new applications. Proc Natl Acad Sci USA 2004; 101 (Suppl. 2): 14593-8.
[http://dx.doi.org/10.1073/pnas.0404887101] [PMID: 15371592]

[34]   Signori A, Gallo F, Bovis F, Di Tullio N, Maietta I, Sormani MP. Long-term impact of interferon or Glatiramer acetate in multiple sclerosis: A systematic review and meta-analysis. Mult Scler Relat Disord 2016; 6(Mar): 57-63.

[http://dx.doi.org/10.1016/j.msard.2016.01.007] [PMID: 27063624]

[35]   Kieseier BC. The mechanism of action of interferon-β in relapsing multiple sclerosis. CNS Drugs 2011; 25(6): 491-502.
[http://dx.doi.org/10.2165/11591110-000000000-00000] [PMID: 21649449]

[36]   Chen M, Valenzuela RM, Dhib-Jalbut S. Glatiramer acetate-reactive T cells produce brain-derived neurotrophic factor. J Neurol Sci 2003; 215(1-2): 37-44.
[http://dx.doi.org/10.1016/S0022-510X(03)00177-1] [PMID: 14568126]

[37]   Blanco Y, Moral EA, Costa M, *et al.* Effect of glatiramer acetate (Copaxone) on the immunophenotypic and cytokine profile and BDNF production in multiple sclerosis: A longitudinal study. Neurosci Lett 2006; 406(3): 270-5.
[http://dx.doi.org/10.1016/j.neulet.2006.07.043] [PMID: 16934924]

[38]   Miller AE. Teriflunomide: a once-daily oral medication for the treatment of relapsing forms of multiple sclerosis. Clin Ther 2015; 37(10): 2366-80.
[http://dx.doi.org/10.1016/j.clinthera.2015.08.003] [PMID: 26365096]

[39]   Marwarha G, Ghribi O. Nuclear Factor Kappa-light-chain-enhancer of Activated B Cells (NF-κB) - a Friend, a Foe, or a Bystander - in the Neurodegenerative Cascade and Pathogenesis of Alzheimer's Disease. CNS Neurol Disord Drug Targets 2017; 16(10): 1050-65.
[http://dx.doi.org/10.2174/1871527316666170725114652] [PMID: 28745240]

[40]   Nicholas JA, Boster AL, Imitola J, O'Connell C, Racke MK. Design of oral agents for the management of multiple sclerosis: benefit and risk assessment for dimethyl fumarate. Drug Des Devel Ther 2014; 8: 897-908.
[PMID: 25045248]

[41]   Gelfand JM, Cree BAC, Hauser SL. Ocrelizumab and Other CD20⁻ B-Cell-Depleting Therapies in Multiple Sclerosis. Neurotherapeutics 2017; 14(4): 835-41.
[http://dx.doi.org/10.1007/s13311-017-0557-4] [PMID: 28695471]

[42]   Dumitrescu L, Constantinescu CS, Tanasescu R. Siponimod for the treatment of secondary progressive multiple sclerosis. Expert Opin Pharmacother 2019; 20(2): 143-50.
[http://dx.doi.org/10.1080/14656566.2018.1551363] [PMID: 30517042]

[43]   Brandstadter R, Katz Sand I. The use of natalizumab for multiple sclerosis. Neuropsychiatr Dis Treat 2017; 13: 1691-702.
[http://dx.doi.org/10.2147/NDT.S114636] [PMID: 28721050]

[44]   Hutchinson M. Natalizumab: A new treatment for relapsing remitting multiple sclerosis. Ther Clin Risk Manag 2007; 3(2): 259-68.
[http://dx.doi.org/10.2147/tcrm.2007.3.2.259] [PMID: 18360634]

[45]   Sellebjerg F, Cadavid D, Steiner D, Villar LM, Reynolds R, Mikol D. Exploring potential mechanisms of action of natalizumab in secondary progressive multiple sclerosis. Ther Adv Neurol Disord 2016; pp. 31-43.

[46]   Ayzenberg I, Hoepner R, Kleiter I. Fingolimod for multiple sclerosis and emerging indications: appropriate patient selection, safety precautions, and special considerations. Ther Clin Risk Manag 2016; 12: 261-72.
[PMID: 26929636]

[47]   Chun J, Hartung HP. Mechanism of action of oral fingolimod (FTY720) in multiple sclerosis. Clin Neuropharmacol 2010; 33(2): 91-101.
[http://dx.doi.org/10.1097/WNF.0b013e3181cbf825] [PMID: 20061941]

[48]   Havrdova E, Horakova D, Kovarova I. Alemtuzumab in the treatment of multiple sclerosis: key clinical trial results and considerations for use. Ther Adv Neurol Disorder 2015; 8(1): 31-45.
[http://dx.doi.org/10.1177/1756285614563522] [PMID: 25584072]

[49]   Ruck T, Bittner S, Wiendl H, Meuth SG. Alemtuzumab in Multiple Sclerosis: Mechanism of Action

and Beyond. Int J Mol Sci 2015; 16(7): 16414-39.
[http://dx.doi.org/10.3390/ijms160716414] [PMID: 26204829]

[50]   Karussis D, Petrou P. Immune reconstitution therapy (IRT) in multiple sclerosis: the rationale. Immunol Res 2018; 66(6): 642-8.
[http://dx.doi.org/10.1007/s12026-018-9032-5] [PMID: 30443887]

[51]   Atakan Z. Cannabis, a complex plant: different compounds and different effects on individuals. Ther Adv Psychopharmacol 2012; 2(6): 241-54.
[http://dx.doi.org/10.1177/2045125312457586] [PMID: 23983983]

[52]   Morgan CJ, Freeman TP, Schafer GL, Curran HV. Cannabidiol attenuates the appetitive effects of Delta 9-tetrahydrocannabinol in humans smoking their chosen cannabis. Neuropsychopharmacology 2010; 35(9): 1879-85.
[http://dx.doi.org/10.1038/npp.2010.58] [PMID: 20428110]

[53]   Niesink RJ, van Laar MW. Does Cannabidiol Protect Against Adverse Psychological Effects of THC? Front Psychiatry 2013 Oct; 16 Oct; pp. 4:130
[http://dx.doi.org/10.3389/fpsyt.2013.00130]

[54]   Colizzi M, McGuire P, Pertwee RG, Bhattacharyya S. Effect of cannabis on glutamate signalling in the brain: A systematic review of human and animal evidence. Neurosci Biobehav Rev 2016; 64: 359-81.
[http://dx.doi.org/10.1016/j.neubiorev.2016.03.010] [PMID: 26987641]

[55]   Di Forti M, Marconi A, Carra E, *et al.* Proportion of patients in south London with first-episode psychosis attributable to use of high potency cannabis: a case-control study. Lancet Psychiatry 2015; 2(3): 233-8.
[http://dx.doi.org/10.1016/S2215-0366(14)00117-5] [PMID: 26359901]

[56]   Schlosburg JE, Kinsey SG, Lichtman AH. Targeting fatty acid amide hydrolase (FAAH) to treat pain and inflammation. AAPS J 2009; 11(1): 39-44.
[http://dx.doi.org/10.1208/s12248-008-9075-y] [PMID: 19184452]

[57]   Alexander JP, Cravatt BF. Mechanism of carbamate inactivation of FAAH: implications for the design of covalent inhibitors and *in vivo* functional probes for enzymes. Chem Biol 2005; 12(11): 1179-87.
[http://dx.doi.org/10.1016/j.chembiol.2005.08.011] [PMID: 16298297]

[58]   Cravatt BF, Demarest K, Patricelli MP, *et al.* Supersensitivity to anandamide and enhanced endogenous cannabinoid signaling in mice lacking fatty acid amide hydrolase. Proc Natl Acad Sci USA 2001; 98(16): 9371-6.
[http://dx.doi.org/10.1073/pnas.161191698] [PMID: 11470906]

[59]   Alonso-Castro AJ, Guzmán-Gutiérrez SL, Betancourt CA, *et al.* Antinociceptive, anti-inflammatory, and central nervous system (CNS) effects of the natural coumarin soulattrolide. Drug Dev Res 2018; 79(7): 332-8.
[http://dx.doi.org/10.1002/ddr.21471] [PMID: 30244493]

[60]   Tsuboi K, Sun YX, Okamoto Y, Araki N, Tonai T, Ueda N. Molecular characterization of N-acylethanolamine-hydrolyzing acid amidase, a novel member of the choloylglycine hydrolase family with structural and functional similarity to acid ceramidase. J Biol Chem 2005; 280(12): 11082-92.
[http://dx.doi.org/10.1074/jbc.M413473200] [PMID: 15655246]

[61]   De Petrocellis L, Cascio MG, Di Marzo V. The endocannabinoid system: a general view and latest additions. Br J Pharmacol 2004; 141(5): 765-74.
[http://dx.doi.org/10.1038/sj.bjp.0705666] [PMID: 14744801]

[62]   Cui N, Wang C, Zhao Z, *et al.* The Roles of Anandamide, Fatty Acid Amide Hydrolase, and Leukemia Inhibitory Factor on the Endometrium during the Implantation Window Front Endocrinol 2017; ; pp. 8:268.

[63]   Fraga D, Zanoni CI, Rae GA, Parada CA, Souza GE. Endogenous cannabinoids induce fever through

the activation of CB1 receptors. Br J Pharmacol 2009; 157(8): 1494-501.
[http://dx.doi.org/10.1111/j.1476-5381.2009.00312.x] [PMID: 19681872]

[64]   Maingret F, Patel AJ, Lazdunski M, Honoré E. The endocannabinoid anandamide is a direct and selective blocker of the background K(+) channel TASK-1. EMBO J 2001; 20(1-2): 47-54.
[http://dx.doi.org/10.1093/emboj/20.1.47] [PMID: 11226154]

[65]   Clapper JR, Moreno-Sanz G, Russo R, *et al.* Anandamide suppresses pain initiation through a peripheral endocannabinoid mechanism. Nat Neurosci 2010; 13(10): 1265-70.
[http://dx.doi.org/10.1038/nn.2632] [PMID: 20852626]

[66]   Ross RA. Anandamide and vanilloid TRPV1 receptors. Br J Pharmacol 2003; 140(5): 790-801.
[http://dx.doi.org/10.1038/sj.bjp.0705467] [PMID: 14517174]

[67]   O'Sullivan SE. Cannabinoids go nuclear: evidence for activation of peroxisome proliferator-activated receptors. Br J Pharmacol 2007; 152(5): 576-82.
[http://dx.doi.org/10.1038/sj.bjp.0707423] [PMID: 17704824]

[68]   Malek N, Popiolek-Barczyk K, Mika J, Przewlocka B, Starowicz K. Anandamide, Acting *via* CB2 Receptors, Alleviates LPS-Induced Neuroinflammation in Rat Primary Microglial Cultures. Neural Plast 2015; 2015130639
[http://dx.doi.org/10.1155/2015/130639] [PMID: 26090232]

[69]   Leung K, Elmes MW, Glaser ST, Deutsch DG, Kaczocha M. Role of FAAH-like anandamide transporter in anandamide inactivation. PLoS One 2013; 8(11)e79355
[http://dx.doi.org/10.1371/journal.pone.0079355] [PMID: 24223930]

[70]   Romano A, Coccurello R, Giacovazzo G, Bedse G, Moles A, Gaetani S. Oleoylethanolamide: a novel potential pharmacological alternative to cannabinoid antagonists for the control of appetite. BioMed Res Int 2014; 2014203425
[http://dx.doi.org/10.1155/2014/203425] [PMID: 24800213]

[71]   Romano A, Karimian Azari E, Tempesta B, *et al.* High dietary fat intake influences the activation of specific hindbrain and hypothalamic nuclei by the satiety factor oleoylethanolamide. Physiol Behav 2014; 136: 55-62.
[http://dx.doi.org/10.1016/j.physbeh.2014.04.039] [PMID: 24802360]

[72]   Basavarajappa BS. Critical enzymes involved in endocannabinoid metabolism. Protein Pept Lett 2007; 14(3): 237-46.
[http://dx.doi.org/10.2174/092986607780090829] [PMID: 17346227]

[73]   Godlewski G, Offertáler L, Wagner JA, Kunos G. Receptors for acylethanolamides-GPR55 and GPR119. Prostaglandins Other Lipid Mediat 2009; 89(3-4): 105-11.
[http://dx.doi.org/10.1016/j.prostaglandins.2009.07.001] [PMID: 19615459]

[74]   Petrosino S, Di Marzo V. The pharmacology of palmitoylethanolamide and first data on the therapeutic efficacy of some of its new formulations. Br J Pharmacol 2017; 174(11): 1349-65.
[http://dx.doi.org/10.1111/bph.13580] [PMID: 27539936]

[75]   Esposito E, Cuzzocrea S. Palmitoylethanolamide is a new possible pharmacological treatment for the inflammation associated with trauma. Mini Rev Med Chem 2013; 13(2): 237-55.
[PMID: 22697514]

[76]   Alhouayek M, Muccioli GG. Harnessing the anti-inflammatory potential of palmitoylethanolamide. Drug Discov Today 2014; 19(10): 1632-9.
[http://dx.doi.org/10.1016/j.drudis.2014.06.007] [PMID: 24952959]

[77]   Hesselink JM, Hekker TA. Therapeutic utility of palmitoylethanolamide in the treatment of neuropathic pain associated with various pathological conditions: a case series. J Pain Res 2012; 5: 437-42.
[http://dx.doi.org/10.2147/JPR.S32143] [PMID: 23166447]

[78]   Orefice NS, Alhouayek M, Carotenuto A, *et al.* Oral Palmitoylethanolamide Treatment Is Associated

with Reduced Cutaneous Adverse Effects of Interferon-β1a and Circulating Proinflammatory Cytokines in Relapsing-Remitting Multiple Sclerosis. Neurotherapeutics 2016; 13(2): 428-38.
[http://dx.doi.org/10.1007/s13311-016-0420-z] [PMID: 26857391]

[79]   Keppel Hesselink JM, Kopsky DJ. Palmitoylethanolamide, a neutraceutical, in nerve compression syndromes: efficacy and safety in sciatic pain and carpal tunnel syndrome. J Pain Res 2015; 8(7): 729-34.
[http://dx.doi.org/10.2147/JPR.S93106] [PMID: 26604814]

[80]   McCarthy DP, Richards MH, Miller SD. Mouse models of multiple sclerosis: experimental autoimmune encephalomyelitis and Theiler's virus-induced demyelinating disease. Methods Mol Biol 2012; 900: 381-401.
[http://dx.doi.org/10.1007/978-1-60761-720-4_19] [PMID: 22933080]

[81]   Martinez NE, Sato F, Omura S, Minagar A, Alexander JS, Tsunoda I. Immunopathological patterns from EAE and Theiler's virus infection: Is multiple sclerosis a homogenous 1-stage or heterogenous 2-stage disease? Pathophysiology 2013; 20(1): 71-84.
[http://dx.doi.org/10.1016/j.pathophys.2012.03.003] [PMID: 22633747]

[82]   Molina-Holgado F, Molina-Holgado E, Guaza C. The endogenous cannabinoid anandamide potentiates interleukin-6 production by astrocytes infected with Theiler's murine encephalomyelitis virus by a receptor-mediated pathway. FEBS Lett 1998; 433(1-2): 139-42.
[http://dx.doi.org/10.1016/S0014-5793(98)00851-5] [PMID: 9738948]

[83]   Mestre L, Iñigo PM, Mecha M, *et al.* Anandamide inhibits Theiler's virus induced VCAM-1 in brain endothelial cells and reduces leukocyte transmigration in a model of blood brain barrier by activation of CB(1) receptors. J Neuroinflammation 2011;; pp. 8:102

[84]   Nagarkatti P, Pandey R, Rieder SA, Hegde VL, Nagarkatti M. Cannabinoids as novel anti-inflammatory drugs. Future Med Chem 2009; 1(7): 1333-49.
[http://dx.doi.org/10.4155/fmc.09.93] [PMID: 20191092]

[85]   Centonze D, Bari M, Rossi S, *et al.* The endocannabinoid system is dysregulated in multiple sclerosis and in experimental autoimmune encephalomyelitis. Brain 2007; 130(Pt 10): 2543-53.
[http://dx.doi.org/10.1093/brain/awm160] [PMID: 17626034]

[86]   Musumeci G, Grasselli G, Rossi S, *et al.* Transient receptor potential vanilloid 1 channels modulate the synaptic effects of TNF-α and of IL-1β in experimental autoimmune encephalomyelitis. Neurobiol Dis 2011; 43(3): 669-77.
[http://dx.doi.org/10.1016/j.nbd.2011.05.018] [PMID: 21672630]

[87]   Eljaschewitsch E, Witting A, Mawrin C, *et al.* The endocannabinoid anandamide protects neurons during CNS inflammation by induction of MKP-1 in microglial cells. Neuron 2006; 49(1): 67-79.
[http://dx.doi.org/10.1016/j.neuron.2005.11.027] [PMID: 16387640]

[88]   Portavella M, Rodriguez-Espinosa N, Galeano P, *et al.* Oleoylethanolamide and Palmitoylethanolamide Protect Cultured Cortical Neurons Against Hypoxia. Cannabis Cannabinoid Res 2018; 3(1): 171-8.
[http://dx.doi.org/10.1089/can.2018.0013] [PMID: 30255158]

[89]   Orio L, Alen F, Pavón FJ, Serrano A, García-Bueno B. Oleoylethanolamide, Neuroinflammation, and Alcohol Abuse Front Mol Neurosci 2019; ; pp.11:490

[90]   Sayd A, Antón M, Alén F, *et al.* Systemic administration of oleoylethanolamide protects from neuroinflammation and anhedonia induced by LPS in rats [published correction appears in Int J Neuropsychopharmacol 2016;; 19(3)

[91]   Gonzalez-Aparicio R, Blanco E, Serrano A, *et al.* The systemic administration of oleoylethanolamide exerts neuroprotection of the nigrostriatal system in experimental Parkinsonism. Int J Neuropsychopharmacol 2014; 17(3): 455-68.
[http://dx.doi.org/10.1017/S1461145713001259] [PMID: 24169105]

[92]   Fu J, Gaetani S, Oveisi F, *et al.* Oleylethanolamide regulates feeding and body weight through activation of the nuclear receptor PPAR-alpha. Nature 2003; 425(6953): 90-3.
[http://dx.doi.org/10.1038/nature01921] [PMID: 12955147]

[93]   Di Filippo M, Pini LA, Pelliccioli GP, Calabresi P, Sarchielli P. Abnormalities in the cerebrospinal fluid levels of endocannabinoids in multiple sclerosis. J Neurol Neurosurg Psychiatry 2008; 79(11): 1224-9.
[http://dx.doi.org/10.1136/jnnp.2007.139071] [PMID: 18535023]

[94]   Baker D, Pryce G, Croxford JL, *et al.* Endocannabinoids control spasticity in a multiple sclerosis model. FASEB J 2001; 15(2): 300-2.
[http://dx.doi.org/10.1096/fj.00-0399fje] [PMID: 11156943]

[95]   Loría F, Petrosino S, Mestre L, *et al.* Study of the regulation of the endocannabinoid system in a virus model of multiple sclerosis reveals a therapeutic effect of palmitoylethanolamide. Eur J Neurosci 2008; 28(4): 633-41.
[http://dx.doi.org/10.1111/j.1460-9568.2008.06377.x] [PMID: 18657182]

[96]   Rahimi A, Faizi M, Talebi F, Noorbakhsh F, Kahrizi F, Naderi N. Interaction between the protective effects of cannabidiol and palmitoylethanolamide in experimental model of multiple sclerosis in C57BL/6 mice. Neuroscience 2015; 290: 279-87.
[http://dx.doi.org/10.1016/j.neuroscience.2015.01.030] [PMID: 25637488]

[97]   Lambert DM, Vandevoorde S, Diependaele G, Govaerts SJ, Robert AR. Anticonvulsant activity of N-palmitoylethanolamide, a putative endocannabinoid, in mice. Epilepsia 2001; 42(3): 321-7.
[http://dx.doi.org/10.1046/j.1528-1157.2001.41499.x] [PMID: 11442148]

[98]   Paterniti I, Impellizzeri D, Di Paola R, Navarra M, Cuzzocrea S, Esposito E. A new co-ultramicronized composite including palmitoylethanolamide and luteolin to prevent neuroinflammation in spinal cord injury. J Neuroinflammation 2013; 10: 91.
[http://dx.doi.org/10.1186/1742-2094-10-91] [PMID: 23880066]

[99]   Contarini G, Franceschini D, Facci L, Barbierato M, Giusti P, Zusso M. A co-ultramicronized palmitoylethanolamide/luteolin composite mitigates clinical score and disease-relevant molecular markers in a mouse model of experimental autoimmune encephalomyelitis. J Neuroinflammation 2019; 16(1): 126.
[http://dx.doi.org/10.1186/s12974-019-1514-4] [PMID: 31221190]

[100] Barbierato M, Facci L, Marinelli C, *et al.* Co-ultramicronized Palmitoylethanolamide/Luteolin Promotes the Maturation of Oligodendrocyte Precursor Cells. Sci Rep 2015; ; 5:16676Published 2015 Nov 18

[101] Skaper SD, Facci L, Fusco M, *et al.* Palmitoylethanolamide, a naturally occurring disease-modifying agent in neuropathic pain. Inflammopharmacology 2014; 22(2): 79-94.
[http://dx.doi.org/10.1007/s10787-013-0191-7] [PMID: 24178954]

[102] Hansen HS. Palmitoylethanolamide and other anandamide congeners. Proposed role in the diseased brain. Exp Neurol 2010; 224(1): 48-55.
[http://dx.doi.org/10.1016/j.expneurol.2010.03.022] [PMID: 20353771]

[103] Paladini A, Fusco M, Cenacchi T, Schievano C, Piroli A, Varrassi G. Palmitoylethanolamide, a special food for medical purposes, in the treatment of chronic pain: a pooled data meta-analysis. Pain Physician 2016; 19(2): 11-24.
[PMID: 26815246]

[104] Schifilliti C, Cucinotta L, Fedele V, Ingegnosi C, Luca S, Leotta C. Micronized Palmitoylethanolamide Reduces the Symptoms of Neuropathic Pain in Diabetic Patients. 2014.
[http://dx.doi.org/10.1155/2014/849623]

[105] Marcucci M, Germini F, Coerezza A, *et al.* Efficacy of ultra-micronized palmitoylethanolamide (um-PEA) in geriatric patients with chronic pain: study protocol for a series of N-of-1 randomized trials

Trials 2016; ; pp. 17:369

[106]   Jean-Gilles L, Feng S, Tench CR, *et al.* Plasma endocannabinoid levels in multiple sclerosis. J Neurol Sci 2009; 287(1-2): 212-5.
[http://dx.doi.org/10.1016/j.jns.2009.07.021] [PMID: 19695579]

CHAPTER 2

Epileptic Seizures Detection Based on Non-linear Characteristics Coupled with Machine Learning Techniques

Firas Sabar Miften[1], Mohammed Diykh[1,4,*], Shahab Abdulla[3] and **Jonathan H. Green[2,3]**

[1] *University of Thi-Qar, College of Education for Pure Science, Nasiriyah, Iraq*

[2] *Faculty of the Humanities, University of the Free State, South Africa*

[3] *Open Access College, University of Southern Queensland, Australia*

[4] *School of Agricultural, Computational and Environmental Sciences, University of Southern Queensland, Australia*

Abstract: The use of transformation techniques (such as a wavelet transform, Fourier transform, or hybrid transform) to detect epileptic seizures by means of EEG signals is not adequate because these signals have a nonstationary and nonlinear nature. This paper reports on the design of a novel technique based, instead, on the domain of graphs. The dimensionality of each single EEG channel is reduced using a segmentation technique, and each EEG channel is then mapped onto an undirected weighted graph. A set of structural and topological graph characteristics is extracted and investigated, and several machine learning techniques are utilized to categorize the graph's attributes. The results demonstrate that the use of graphs improves the quality of epileptic seizure detection. The proposed method can identify EEG abnormities that are difficult to detect accurately using other transformation techniques, especially when dealing with EEG big data.

Keywords: Epileptic EEG Signals, Graphs, Modularity, Multi-Channel, Statistical Features.

INTRODUCTION

Epilepsy is a chronic neurologic disorder characterized by recurrent unprovoked seizures. The disorder affects more than 40 million people around the world, most of them young children and older adults. Developing countries contribute around 85% of those epilepsy cases. Several clinical studies have shown that epilepsy can

* **Corresponding author Mohammed Diykh:** University of Southern Queensland;University of Thi-Qar, College of Education for Pure Science, Department of Computer science; E-mail: Mohammed.diykh@usq.edu.au

be caused by interactions between several genes and environmental factors. However, people who have suffered brain damage, strokes, toxicity, and high fever could be prone to epileptic seizures. During an epileptic seizure, the brain neurons discharge suddenly, and abnormal activities occur within the cerebral cortex.

A seizure usually lasts about two minutes or less, depending on the age of the patient and the health of the brain [1]. Epileptic seizures can be controlled by medications and, in some severe cases, surgery could be a solution. To diagnose epilepsy, an electroencephalogram (EEG) as a means of recording the electrical activity of the brain and how brain neurons are functioning can detect abnormalities in brain activity.

Much research has been done based on using EEG signals to identify and trace the abnormities produced by epileptic seizures. Many of these studies have been conducted based on wavelet and Fourier transformations [2 - 5]. For example, Kumar *et al*. [4] analysed epileptic seizures by adopting a wavelet transformation based on an approximate entropy. An EEG signal was decomposed into five levels. Then, the approximate entropy was calculated as a representative feature. Two machine learning algorithms—a support vector machine (SVM) and a neural network—then were utilised to classify EEG features into epileptic and non-epileptic segments. Orosco *et al*. [5] employed a stationary wavelet transform to extract EEG features. Several spectral features were extracted and investigated using a stepwise approach to identify the most effective feature set for detecting epileptic seizures. In another study, Bajaj and Pachori [1] applied an empirical mode decomposition to study the behaviour of EEG signals during epileptic seizures: the characteristics of bandwidth were used to classify EEG signals into epileptic and non-epileptic using an SVM. Moreover, Hassan *et al*. [3], in considering 6 spectral features, showed that a tunable-Q factor wavelet transform (TQWT) with a bagging technique was a good tool for detecting epileptic seizures. A combination of several machine learning algorithms was integrated as an ensemble classifier. Guler *et al*. [2] studied wavelet coefficients with the Lyapunov exponent to analyse EEG signals: in this case, an SVM classifier was utilised to differentiate EEG groups. All the above-mentioned research was conducted using the Bonn University data (described below) that is used in our study.

More recently, Gao *et al*. [6] suggested a recurrence time-based approach to detect epileptic EEG signals. In that study, EEG signals were segmented into overlap and non-overlap intervals. Three EEG channels were involved to record epileptic EEG signals. Another study was conducted by Gao *et al*. [7], in which a nonlinear adaptive multiscale decomposition approach was proposed to analyse nonstationary signals. Epileptic EEG signals were taken as an example in that study. Gao *et al.* [8] also studied epileptic EEG signals using several complex measures. Amin *et al.* [9] utilised a wavelet transform to decompose EEG signals into approximations and detail coefficients. In that study, an arithmetic coding model was used to convert wavelet coefficients into bitstreams. San-Segundo *et al.* [10] suggested a deep neural network model to detect seizures in EEG signals. In that study, they employed two convolutional layers for the feature extraction and three layers for the classification. Wang *et al.* [11] applied Fourier transform to analyse EEG signals. A set of features was extracted and reduced using a principal component analysis. A random forest model coupled with a grid search optimization was utilised to classify the extracted features.

This paper describes an approach that addresses the detection of epileptic seizures in EEG signals in a way that differs from that of preceding studies. In this study, the dimensionality in EEG signal data is reduced to extract relevant information before mapping it to graphs, simultaneously eliminating data that is redundant. The statistical characteristics of EEG signals are extracted through a segmentation technique. Based on our previous work, using statistical characteristics of EEG signals in constructing graphs could contribute to improving the ability of the graphs to reflect any abnormal behaviours in EEG signals [12]. In this method, after a thorough investigation, we segment each single-channel EEG signal into four windows of 1024, 1024, 1024, and 1025; then, each segment is split further into 32 clusters. Eight statistical characteristics are extracted from each cluster, and a vector of statistical features is then mapped into an undirected graph. To detect the abnormal patterns of epileptic seizures in EEG signals, Jaccard coefficients, degree distribution, modularity, entropy, high degree, low degree, and other local and global graph characteristics are extracted and analysed. Different machine learning techniques, including least support vector machine (LS-SVM), Naïve Payson (NB), k-means, and k-nearest, are used to categorize network characteristics into different EEG cases. The main objective of this method is to improve the detection accuracy of epileptic seizures in EEG signals by integrating the graph concept with a statistical model, thus determining the best graph feature set for analysing epileptic EEG signals.

EEG DATA

The epileptic EEG datasets used in this research are available online from http://epileptologie -bonn.de/ cms/front_content.php?idcat=193 &lang=3. The datasets are considered a clinical EEG benchmark database for most epileptic research. Further details of the datasets are described [13].

PROPOSED METHODOLOGY

The proposed methodology presented in Fig. (**1**) consists of four steps. Firstly, each single EEG signal is segmented into intervals to reduce its dimensionality and eliminate redundant information. Then, each EEG channel is mapped to an undirected graph. In this study, we use the same procedure as in our previous work [12] to transfer EEG channels into graphs, in which each statistical feature is considered a node. The connections among graph nodes are made in accordance with a procedure used by Zhang and Small [14].

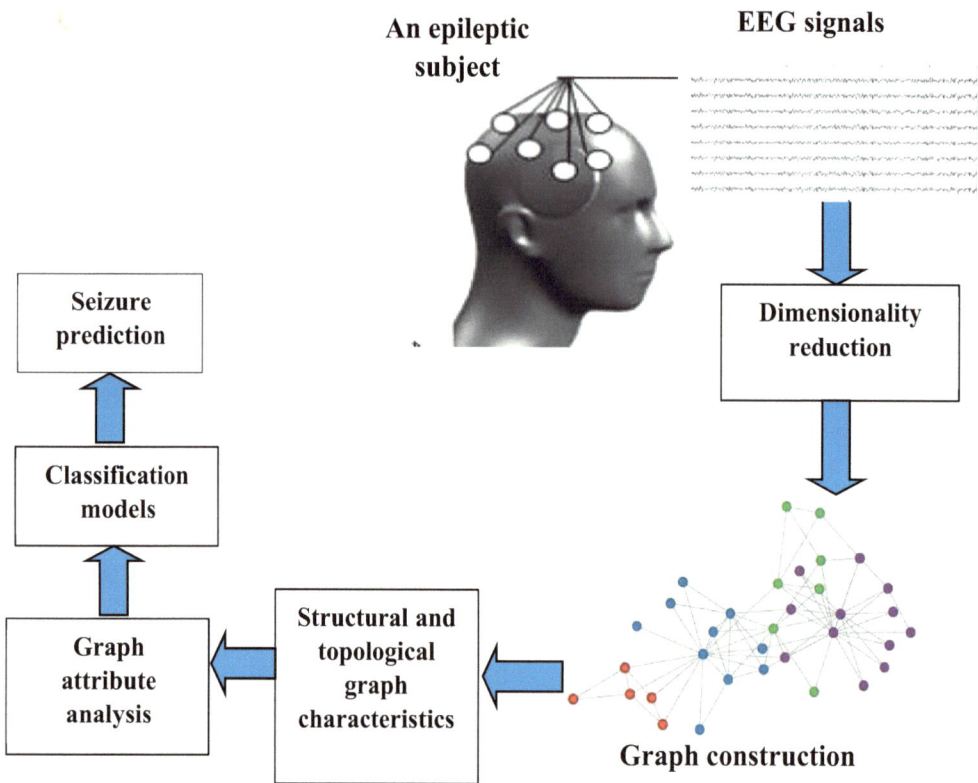

Fig. (1). Proposed methodology to detect epileptic seizures.

The weak nodes in each graph are eliminated using a predefined threshold because they have a negative impact on the behaviour of the graph and, as the analysis showed, cause misclassification. A set of structural and topological graph characteristics are examined, and the extracted graph characteristics then fed to the various machine learning algorithms to detect epileptic seizures.

EEG DIMENSIONALITY REDUCTION

Our previous studies demonstrated that partitioning EEG signals into small intervals to reduce the dimensionality provided adequate EEG sleep classification results. In this study, a segmentation technique is used to eradicate the redundant EEG data and retain the most important information. Each single-channel EEG signal is partitioned into four intervals, 1024, 1024, 1024, and 1025, respectively. Then, each interval is further divided into twelve blocks; this number of blocks is chosen empirically after a thorough investigation. As a result, each EEG channel is segmented into 128 blocks, and a set of representative features is then extracted from those 128 blocks. Eight statistical features are drawn from each block to obtain a vector of 1024 features to represent each EEG channel. The eight features are {median, maximum, minimum, mode, range, standard deviation, variation, kurtosis} and they are denoted as $\{ X_{Me}, X_{Max}, X_{Min}, X_{Mod}, X_{Rand}, X_{SD}, X_{Var}, X_{Kue} \}$.

MAPPING STATISTICAL FEATURES TO GRAPHS

Previous neurologic studies have shown graph theory to be a useful tool in investigating human brain behaviour. In particular, it can be used to study the dynamic characteristics of most brain signals. However, the topological and structural characteristics of graphs and how they change during transitions across epileptic EEG signals are still unknown. In this study, it is demonstrated that epileptic seizures can be detected through the behaviours of the graph characteristics.

Let us consider a set of statistical features, denoted by $L=\{l_1,l_2,x_3,\ldots\ldots,l_n\}$, with each data point representing a feature. In accordance with our previous study [12], vector L is transferred into a graph. In the present study, each feature in L is considered as a node in graph G. If two nodes vi and vj are connected, their edge is calculated based on Euclidean distance.

To eradicate the nodes with poor connections, a predefined threshold (ä) is adopted. Each connection that is lower or equal to this threshold is eliminated.

$$(v_i, v_j) \in E, if \ d(v_i, v_j) \leq \delta \qquad (1)$$

To describe the connections of the network nodes, an adjacency matrix A of graph G is calculated. The adjacent matrix of an undirected graph is symmetrical, *i.e.*, $A(v_1,v_2) = A(v_2,v_1)$

One interesting observation from this investigation was that the use of a combination of topological and structural graph attributes could reflect highly discernible abnormalities in EEG signals during epileptic episodes.

TOPOLOGICAL AND STRUCTURAL GRAPH PROPERTIES

The graph characteristics that follow are considered in epileptic seizure detection in this study [13 - 17].

1. Jaccard Coefficient

The Jaccard coefficient is a statistical approach used for discovering the differences and similarities between a graph's nodes. It computes by dividing the set of the intersection nodes by the set of the union nodes between any nodal couple in the graph. The Jaccard coefficient is defined using the following equation:

$$w(v_1, v_2) = \frac{|\Gamma(v_i) \cap \Gamma(v_j)|}{|\Gamma(v_i) \cup \Gamma(v_j)|} \tag{2}$$

where $\Gamma(vi)$ refers to the set of neighbours of node v_j, $\Gamma(vj)$ refers to the set of neighbours of vj, and $w = [0, 1]$

2. Degree Distribution

The degree distribution for a network is calculated using the following formula:

$$P(k) = \frac{|\{v|d(v) = k\}|}{N} \tag{3}$$

where $d(v)$ is the degree of node v, and v is the number of nodes in the graph

3. Modularity

Modularity is a mathematical approach that is used to measure the strength of dividing a graph into groups of nodes. Each group in a graph must have a high relationship among its nodes and lower connectivity towards other groups. This graph feature is utilised in most social networks to identify communities.

Let $A_{i,j}$ be an adjacency matrix of a network G. The mathematical formula of modularity is defined as

$$Q = \frac{1}{2m} \sum_{i,j} (A_{i,j} - \frac{k_i k_j}{2m}) \delta(C_i C_j) \tag{4}$$

where $m = 1/2 \sum_i k_i$ denotes the total number of the links, and k_i refers to the

degree of node i and C_i is the cluster name of node i . δ $(c_i c_i)$ takes value 1 if two nodes i,j belong to the same cluster; otherwise, it is equal 0.

4. Average Degree of the Graph

This is the number of all the nodes connected to node v_i . The following equation is the mathematic formula of node degree:

$$d_i = \sum_{j \in \prod(I)} \tag{5}$$

where j represents all the connected nodes to v_i. However, to calculate the degree of the entire graph, the following formula is used:

$$D = \sum_i d_i \tag{6}$$

5. Closeness Centrality

Closeness Centrality reveals the important node in a graph that connects with other nodes quickly in a graph. The closeness centrality is defined as:

$$C_{clos} = \frac{1}{\sum_{i \in V} dist(v_i, v_j)} \tag{7}$$

where *dist* (v_i, v_j) indicates the shortest path between nodes v_i and v_j .

6. Clustering Coefficients (*CC*)

The clustering coefficients (*CC*) are defined as the fraction of the connected nodes to node v_i . The main formula of clustering coefficients is defined as

$$CC_{v_i} = \frac{\frac{m1}{m2}}{2} \tag{8}$$

where $m1$ refers to the number of the actual links between node v_i with its neighbours, and $m2$ refers to the number of the neighbours of v_i .

MACHINE LEARNING TECHNIQUES

Four machine learning algorithms—namely, a least support vector machine, k-means, k-nearest, and naïve Bayes—were used to identify epileptic seizures in EEG signals in this study. The parameters of these algorithms were empirically determined.

1. Least Square Support Vector Machine (LS-SVM)

The LS-SVM was used to classify epileptic EEG categories. The graph features were extracted and then they were used as the input to the LS-SVM. Suppose a

training set $\{ X_i, Y_i \}$ where $i = 1,2,.... .,N$, x_i is the input features set, and $y_i \in \{1, -1\}$ is the class label of x_i. The main function of LS-SVM can be described as

$$y(x) = sign[w^t \emptyset(x) + b] \qquad (9)$$

Where w is the weight set, b is the bias. In this paper, we referred to the nonlinear function as $\emptyset$. To obtain high classification results, w and b need to be determined. We utilised an optimization function to obtain w and b. In addition, we employed the radial basis function as the kernel function in this research. The performance of the LS-SVM classifier is evaluated using sensitivity, specificity and accuracy.

2. Naive Bayes (NB)

NB is a statistical classifier that is commonly employed to predict the probabilities of class memberships. In biomedical research, NB has been proved to yield high accuracy and low computation time with large datasets. Suppose $Z = \{ z_1, z_2,....,z_n \}$ is a feature set that contains n features. Let H denote to a hypothesis of each set of Z which belongs to a specific group. Based on Naive Bayes rules, Z is considered as evidence, and it seeks to assign each point of Z to the highest posterior probability class. Naive Bayes is also adopted in this research to categorise the attributes of the graphs.

3. k-Nearest Classifier

Because it is simple, adaptive in nature, and robust with noisy datasets, most EEG research has used k-nearest as a classifier. It uses local information to predict unknown data. First, the algorithm finds the number of nearest neighbours and then it classifies the input data into several classes using the first step. To find the nearest neighbour, the Euclidean distance was used in this study.

4. k-Means

K-means is one of the most effective unsupervised classification algorithms. It has been used to solve multi clustering problems It separates a given dataset into several groups based on k centroids for each group. The algorithm works according to the following steps.

1. Select k data points as the cluster centres.

2. Generate the membership matrix u according to the following equation.

$$u = \begin{cases} 1 \ if \ ||x_j - c_i||^2 \leq ||x_j - c_t||^2 \\ 0 \qquad otherwise \end{cases} \tag{10}$$

3. Calculate the objective function j using the following equation

$$j = \sum_{j=1}^{k} (\sum_{k} ||x_k - c_i||^2) \tag{11}$$

4. Update the cluster centres based on the following equation.

$$c_i = \frac{\sum_{j=1}^{n} u_{i,j} x_j}{\sum_{j=1}^{n} u_{i,j}} \tag{12}$$

5. Go to step 2 while the cluster centres are moving, otherwise, go to step 6.

6. End.

RESULTS AND DISCUSSION

Analysing Graph Attributes

The topological and structural graph attributes—namely, clustering coefficients, degree distribution, modularity, clustering coefficients, closeness centrality, entropy, and average degree—were extracted to recognize epileptic EEG signals. It was noticed that the use of a combination of structural and topological graph characteristics reflected the sudden abnormalities in EEG signals during an epileptic seizure. The four machine learning algorithms were trained with different combinations of parameters. Several experiments were conducted to obtain the optimal parameters for each algorithm. At each experiment, the obtained results were recorded as well as the paraments. Once the classification results were improved, the parameters were fixed. The proposed model was conducted on different pairs of EEG signals. Based on our results, the best classification results were obtained for (A *vs* E), (B *vs* E), (D *vs* E), (C *vs* E), ((A, B) *vs* E), ((C, D) *vs* E), and ((A, C and D) *vs* E) when $\gamma = 10$ and $\sigma 2 = 4$ for the LS-SVM, k=2 for k-means, k=7 for k-nearest. All experiments are conducted using MATLAB software.

Fig. (**2**) shows an example of how the box plots of the four topological and structural graph characteristics were used to investigate whether the graph attributes could recognise EEG groups. The Jaccard coefficient, average degree, clustering coefficient, and closeness centrality were used as examples in this figure. As can be observed in Fig. (**2**), the Jaccard coefficients of the graphs showed different values, ranging from 0 to 1 for each EEG case. Groups D and E gained the highest values of all the cases; thus, detecting EEG cases was better facilitated by using the Jaccard coefficients.

Clustering coefficients were also used to investigate their ability to characterize the EEG cases. The obtained results showed that the clustering coefficients could reflect the changes in EEG signals during each case. The values of clustering coefficients lay between [0, 1]. As presented in Fig. (**2**), Set C gained the highest clustering coefficients among the five EEG cases.; however, Group E had the lowest values of clustering coefficients among the EEG cases. The results demonstrated, thus, that the clustering coefficients had the capability to differentiate the EEG groups.

The average degree of the graphs was also investigated and showed a positive reflection to detect the epileptic seizures in EEG signals. It can be observed that each EEG group has different values of the average degree. For example, the value of the average degree of Group E ranged between 360-460—the highest set of values among the five groups. However, Group D was recorded as having the lowest set of values of the average degree. The results demonstrated that the average degree of the graphs could be used to detect seizure activity in EEG signals. The closeness attribute was also investigated and it performed well in identifying Groups E and D. However, it did not perform well in distinguishing between Groups A and B, as the two groups had very close characteristics.

In this study, the Mann-Whitney U Test was conducted to identify the ability of the graph attributes to recognize different EEG categories. The attributes of the graph of each pair of EEG categories including: (A *vs* E), (B *vs* E), (D *vs* E), (C *vs* E), ((A, B) *vs* E), ((C, D) *vs* E), and ((A, C and D) *vs* E), were investigated and tested. The results of the tests are reported in Table **1**. It was found that all the characteristics of the graphs differed significantly (p<0.05). In addition, it was found that there was a large difference among the five EEG groups. Thus, the Mann-Whitney U Test showed that the proposed method had high potential to differentiate the individual EEG Groups of A-E, as well as the combined EEG groups against Group E. To evaluate the performance of the proposed methodology, several experiments were conducted. The graph features were assessed independently to determine the ability of each graph attribute in analysing EEG data.

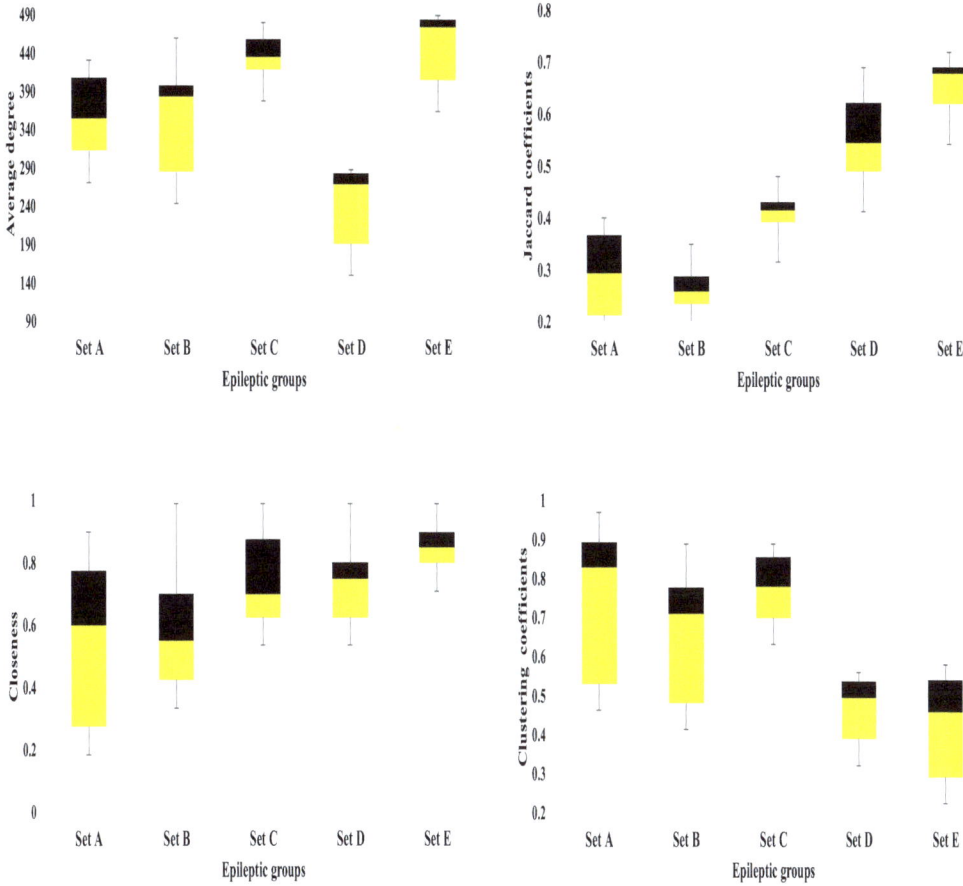

Fig. (2). Boxplots of the graph attributes.

EEG groups of A-E, as well as using a combination of EEG groups against group E.

Table 1. Mann-Whitney U Test of The Networks Characteristics for Each Pair of EEG Cases.

p-value	A VS E	B VS E	C VS E	D VS E	(A and B) *vs* E	(C and D) *vs* E	(A,C and D) *vs* E
Jaccard coefficients	0.70e-12	7.421e-01	1.63e-11	0.00012	1.98 e-13	0.00012	2.13e-10
Clustering coefficients	3.21e-14	1.64e-12	0.0024	0.0014	1.89e13	3.35e-14	0.00036

(Table 1) cont.....

p-value	A VS E	B VS E	C VS E	D VS E	(A and B) *vs* E	(C and D) *vs* E	(A,C and D) *vs* E
Closeness	0.00456	0.00140	6.52e-12	0.00640	0.00780	2.97e-19	0.0078
Average degree	4.21e-17	3.78e-19	0.0054	3.21e-19	0.0098	1.98-e210	1.12e-114
Modularity	2.20e-16	5.201e-03	1.25e-15	0.0012	0.0035	0.0067	4.98e-10
Degree distribution	0.00023	1.45e-14	3.12e-13	0.00064	2.12e-19	1.65-18	0.00071
Entropy	0.00021	0.000010	0.00052	0.000090	0.00011	0.00039	0.00048

Classification Results

The LS-SVM was utilised to classify the graph characteristics with the following parameters: the RBF kernel with ã=1 and ó=1. The EEG dataset was divided into training and testing groups. Fig. (3) shows the accuracy of the proposed method based on each graph feature. One of the noticeable observations in this study was that the topological graph attributes represented by entropy clustering coefficients, degree distribution, Jaccard coefficients, and modularity yielded a high level of accuracy compared to other graph characteristics. These attributes had the highest discriminative ability among the graphs features.

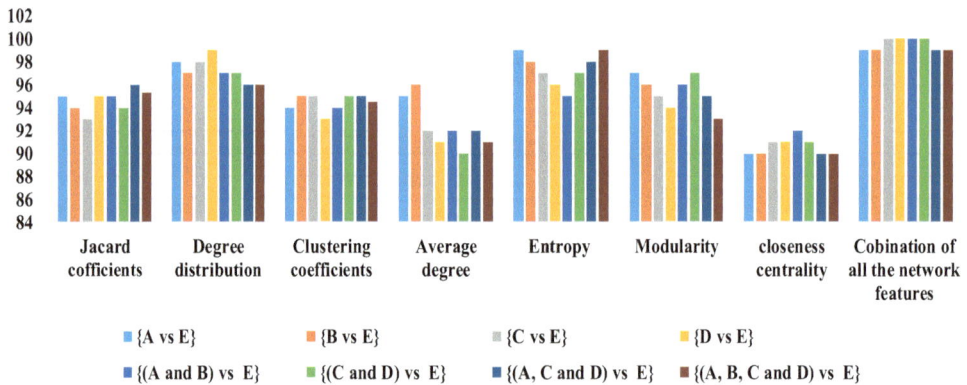

Fig. (3). Detection EEG cases based on graph.

Average accuracy of 99% was obtained using the topological graph attributes to detect all the EEG cases. Our findings also showed that using a combination of topological and structural graph attributes yielded a higher detection rate than using only a single graph attribute. The proposed method was evaluated using different statistical measures including sensitivity, accuracy, and specificity. Table

2 reports the classification results for all the EEG cases using the four machine learning techniques. The mean and standard deviation of sensitivity, accuracy and specificity were calculated for all EEG cases. We considered the combination of the graph characteristics as the key features in all the experiments to identify EEG cases. The results in Table **2** show that the proposed method using the LS-SVM yielded the highest accuracy across all the cases; however, the k-means achieved an average accuracy of 98% across the cases, and it was very close to those obtained by the LS-SVM. It had the second-highest classification accuracy in detecting epileptic seizures in EEG signals. The NB classifier slightly lower than that of the k-means and the LS-SVM—was ranked as obtaining the third highest accuracy, sensitivity and specificity. The results demonstrated that, for k-means and NB, the proposed method had very similar results across all cases.

The k-nearest classier was also used in this paper. Different values of k were tested in this experiment and the results demonstrated that k=10 gave a higher classification accuracy than k=3 and k=6. The k-nearest results were relatively similar to those by the k-means, NB, and the LS-SVM. Thus, by comparing the classification results of the four classifiers, we could determine that the LS-SVM was the most accurate classifier in categorizing the features of the graphs.

Classification model	Accuracy		Specificity		Sensitivity	
	Mean	Median	Mean	Median	Mean	Median
LS-SVM	99	98	97.1	97	98.9	99
k-means	97.5	96.8	96.2	96	96	97
NB	94.1	94	93.4	93	93.8	94
k-nearest	94	94	93	92.8	92	92.7

CONCLUSION

Accurate detection of epilepsy (and capturing persistent features related to this disease) with a fast, automated and a robust modelling approach remains a significant challenge in the medical area. To address this challenge, this paper proposes an efficient epileptic detection technique combined with a statistical approach that can be used for data dimensionality reduction and feature extraction to yield a reliable and versatile classification model applied in the detection of epileptic conditions. In sum, to generate a reliable EEG classification model, the EEG signals were partitioned into segments; then, a set of statistical features was extracted from each cluster to form a vector of features; each vector was mapped into an undirected graph; and, finally, several topological and structural graphs attributes were extracted and investigated. Our findings showed that topological graph attributes, such as entropy, had the potential to show abnormalities in EEG signals during epileptic seizures.

While the present study has led, demonstrably, to an improved method for the detection of epilepsy, an independent study in the future could also apply the proposed technique to detect many other issues, such as signs of sleep spindles and k-complexes prevalent in EEG signals. We contend that the proposed graph-based machine learnings method can be explored further to develop a seizure warning system within a medical diagnostic platform that can potentially assist practising neurologists to better diagnose and treat neurological disorders based on EEG signals.

CONSENT FOR PUBLICATION

Not applicable.

CONFLICT OF INTEREST

The authors declare that no conflict of interest.

ACKNOWLEDGEMENTS

The authors acknowledge EEG data from the University of Bonn.

REFERENCES

[1]   Bajaj V, Pachori RB. Classification of seizure and non-seizure EEG signals using empirical mode decomposition. IEEE Trans Inf Technol Biomed 2012; 16(6): 1135-42.
      [http://dx.doi.org/10.1109/TITB.2011.2181403] [PMID: 22203720]

[2]   Güler I, Ubeyli ED. Multiclass support vector machines for EEG-signals classification. IEEE Trans Inf Technol Biomed 2007; 11(2): 117-26.
      [http://dx.doi.org/10.1109/TITB.2006.879600] [PMID: 17390982]

[3]   Hassan AR, Siuly S, Zhang Y. Epileptic seizure detection in EEG signals using tunable-Q factor wavelet transform and bootstrap aggregating. Comput Methods Programs Biomed 2016; 137: 247-59.
      [http://dx.doi.org/10.1016/j.cmpb.2016.09.008] [PMID: 28110729]

[4]   Kumar Y, Dewal M, Anand R. Epileptic seizures detection in EEG using DWT-based ApEn and artificial neural network. Signal Image Video Process 2014; 8: 1323-34.
      [http://dx.doi.org/10.1007/s11760-012-0362-9]

[5]   Orosco L, Correa AG, Diez P, Laciar E. Patient non-specific algorithm for seizures detection in scalp EEG. Comput Biol Med 2016; 71: 128-34.
      [http://dx.doi.org/10.1016/j.compbiomed.2016.02.016] [PMID: 26945460]

[6]   Gao J, Hu J. Fast monitoring of epileptic seizures using recurrence time statistics of electroencephalography. Front Comput Neurosci 2013; 7: 122.
      [http://dx.doi.org/10.3389/fncom.2013.00122] [PMID: 24137126]

[7]   Gao J, Hu J, Tung W-w. Facilitating joint chaos and fractal analysis of biosignals through nonlinear adaptive filtering
      [http://dx.doi.org/10.1115/DSCC2011-6083]

[8]   Gao J, Hu J, Tung WW. Complexity measures of brain wave dynamics. Cogn Neurodyn 2011; 5(2):

171-82.
[http://dx.doi.org/10.1007/s11571-011-9151-3] [PMID: 22654989]

[9]   Amin HU, Yusoff MZ, Ahmad RF. A novel approach based on wavelet analysis and arithmetic coding for automated detection and diagnosis of epileptic seizure in EEG signals using machine learning techniques. Biomed Signal Process Control 2020; 56: 777-80.
[http://dx.doi.org/10.1016/j.bspc.2019.101707]

[10]   San-Segundo R, Gil-Martín M, D'Haro-Enríquez LF, Pardo JM. Classification of epileptic EEG recordings using signal transforms and convolutional neural networks. Comput Biol Med 2019; 109: 148-58.
[http://dx.doi.org/10.1016/j.compbiomed.2019.04.031] [PMID: 31055181]

[11]   Wang X, Gong G, Li N, Qiu S. Detection Analysis of Epileptic EEG Using a Novel Random Forest Model Combined With Grid Search Optimization. Front Hum Neurosci 2019; 13: 52.
[http://dx.doi.org/10.3389/fnhum.2019.00052] [PMID: 30846934]

[12]   Diykh M, Li Y, Wen P. EEG sleep stages classification based on time domain features and structural graph similarity. IEEE Trans Neural Syst Rehabil Eng 2016; 24(11): 1159-68.
[http://dx.doi.org/10.1109/TNSRE.2016.2552539] [PMID: 27101613]

[13]   Andrzejak RG, Lehnertz K, Mormann F, Rieke C, David P, Elger CE. Indications of nonlinear deterministic and finite-dimensional structures in time series of brain electrical activity: dependence on recording region and brain state. Phys Rev E Stat Nonlin Soft Matter Phys 2001; 64(6 Pt 1)061907
[http://dx.doi.org/10.1103/PhysRevE.64.061907] [PMID: 11736210]

[14]   Zhang J, Small M. Complex network from pseudoperiodic time series: topology versus dynamics. Phys Rev Lett 2006; 96(23)238701
[http://dx.doi.org/10.1103/PhysRevLett.96.238701] [PMID: 16803415]

[15]   Blondel VD, Guillaume J-L, Lambiotte R, Lefebvre E. Fast unfolding of communities in large networks. J Stat Mech 2008; 2008P10008
[http://dx.doi.org/10.1088/1742-5468/2008/10/P10008]

[16]   Clauset A, Newman ME, Moore C. Finding community structure in very large networks. Phys Rev E Stat Nonlin Soft Matter Phys 2004; 70(6 Pt 2)066111
[http://dx.doi.org/10.1103/PhysRevE.70.066111] [PMID: 15697438]

[17]   Fu K, Qu J, Chai Y, Dong Y. Classification of seizure based on the time-frequency image of EEG signals using HHT and SVM. Biomed Signal Process Control 2014; 13: 15-22.
[http://dx.doi.org/10.1016/j.bspc.2014.03.007]

CHAPTER 3

Hampering Essential Tremor Neurodegeneration in Essential Tremor: Present and Future Directions

Rania Aro[1], **Pierre Duez**[1,*], **Amandine Nachtergael**[1] and **Mario Manto**[2,*]

[1] *Unit of Therapeutic Chemistry and Pharmacognosy, University of Mons (UMONS), Belgium*

[2] *Department of Neurology, CHU-Charleroi, Chaussée de Bruxelles 140, 6042 Lodelinsart, Belgium*

Abstract: Essential tremor (ET) is one of the most prevalent neurological disorders worldwide. ET presents mainly with kinetic and action tremor in upper limbs. Tremor may also affect the head and some patients develop an ataxic gait, as well as cognitive/affective symptoms. ET significantly impacts the quality of life. There is accumulating evidence that ET is a slowly progressive neurodegenerative disease, driven by both genetic and environmental (possibly dietary) factors. Both the olivo-cerebellar pathways and the cerebellar cortex are critically involved, with particular impairments in the morphology of the Purkinje neurons (Purkinjopathy) as well as the surrounding micro-circuitry. Dysfunctional cerebello-thalamo-cortical loops probably result in bursts of tremor. So far, only few symptomatic medications are available, including beta-blockers, primidone and drugs aiming to modulate GABAergic transmission such as topiramate or gabapentine. Surgery (deep brain stimulation, thalamotomy) is proposed to refractory cases but carries the risk of infection, bleeding in the brain and several technical issues related to the mispositioning of electrodes. MRI-guided focused ultrasound is a promising technique, but long-term follow-up is missing. Repetitive transcranial magnetic stimulation (rTMS) and transcranial direct current stimulation (tDCS) are encouraging non-invasive techniques but no consensus on optimal protocols has been reached so far. It is remarkable to observe that none of the available therapies targets the neurodegenerative process affecting in particular the cerebellum, the masterpiece of progression of the disease. This chapter focuses on the pathogenesis of ET and discusses possible novel avenues for therapy and prevention. In particular, the impact of environmental toxins such as beta-carboline alkaloids (βCAs), possibly generated from Maillard-type reaction products, is discussed. Animal models of ET, toxicokinetics and neurotoxic effects of βCAs are presented, with an emphasis on the neuroprotective pathways that are candidates to block the neurodegenerative process. Moreover, we consider a group of enzymes that could be neuroprotective, especially GAD65 and GAD67, involved in GABA synthesis/neurotransmission, and MAO_A/MAO_B. Finally, we emphasize the potential interest of dietary phytochemicals

* **Corresponding authors Pierre Duez:** Unit of Therapeutic Chemistry and Pharmacognosy, University of Mons (UMONS), Bldg 6, 25 Chemin du champ de Mars, 7000 Mons, Belgium; Tel: +3265373509; Fax: +3265373351; E-mail: Pierre.DUEZ@umons.ac.be **Mario Manto:** Department of Neurology, CHU-Charleroi, Chaussée de Bruxelles 140, 6042 Lodelinsart, Belgiu; Tel/Fax : +3271921311; E-mail: mmanto@ulb.ac.be

Atta-ur-Rahman & Zareen Amtul (Eds.)

(such as phenolic acids, catechins, flavonoids, anthocyans, stilbenoids, curcuminoids) and herbal therapies (based *i.e.* on *Bacopa monnieri, Ginkgo biloba*) as neuroprotective approaches to hamper the neurodegenerative process in ET.

Keywords: β-Carboline Alkaloids, Chemoprevention, CNS Disorder, Deep Brain Stimulation of the Thalamus, Essential Tremor (ET), ET Pharmacotherapy, ET Animal Model, (GABA)ergic Dysfunction, Gamma Knife Surgery (GK), Harmane, Harmaline, Maillard Reaction, Neuroprotection, Neurodegenerative, Neurotoxicity, Purkinje Neurons, Repetitive Transcranial Magnetic Stimulation (rTMS), Thalamotomy.

ESSENTIAL TREMOR: DEFINITION AND EPIDEMIOLOGY

Essential tremor (ET) is characterized by a slowly progressive postural and/or kinetic involuntary tremor, a bilateral action tremor affecting predominantly the arms, the head and/or the voice [1]. ET is primarily a kinetic tremor; the main clinical features of ET consist in kinetic tremor of the arms (tremor occurring during guided voluntary movements) with frequencies of 4 to 12 Hz, followed by postural and/or kinetic tremor of cranial structures (*i.e.* neck, jaw, voice) [2]. Patients usually first become aware of the tremor when they are holding newspaper or utensils or when reaching for objects. When ET affects the neck muscles, patients exhibit either yes-yes or no-no oscillations of the head. Furthermore, ET can affect the vocal cords, causing a tremulous voice while singing or talking [3]. As time evolves, ET tends to impair balance and gait, and may even cause falls.

Apart from the first group of motor features, recent research points out a variety of cognitive and psychiatric signs. One of the most common non-motor symptoms in ET is the presence of mild cognitive deficits [4], notably for verbal fluency, naming, mental set-shifting, verbal memory, and working memory; deficits in olfaction and hearing loss have also been observed in ET but the published studies remain inconclusive [5]. Significant relationships are reported between ET and depression [5], poor nocturnal sleep quality and sleep disturbances [6]. Non-motor symptoms could be a part of the disease in the early stages; indeed depression and anxiety are more common in young patients with ET [6]. In fact, depressive symptoms appear to be stronger predictors of tremor-lowered quality of life than the motor aspects of tremor itself [7].

Obviously, and even if some patients will never come to medical attention, both motor and non-motor ET symptoms result in significant psychosocial and physical disabilities, interfering with activities of daily living (ADL) such as eating, drinking, writing [3]. The classical view of ET as a monosymptomatic condition is now replaced by the concept of a heterogeneous disorder with multiple motor and non-motor features of varying degrees.

Incidence and Prevalence

ET is among the most prevalent disabling and poorly understood neurological movement disorders, especially affecting elderly people, but also appearing in young adults and even during childhood [8]. The disease has been reported not only by neurologists, but also by internists, geriatricians, and general practitioners [9]. The adjusted incidence is about 619 per 100.000 person-years among persons aged 65 and older [10]. However, the prevalence estimates have varied enormously amongst studies and it is therefore difficult to establish the prevalence at a world level. A meta-analysis by Louis and Ferreira identified 28 studies over 19 countries, with prevalence ranging from 0.01% (Nigeria, China; all ages) to 20.5% (USA; over 65 years) (Fig. **1**). By pooling prevalence in all age classes, the

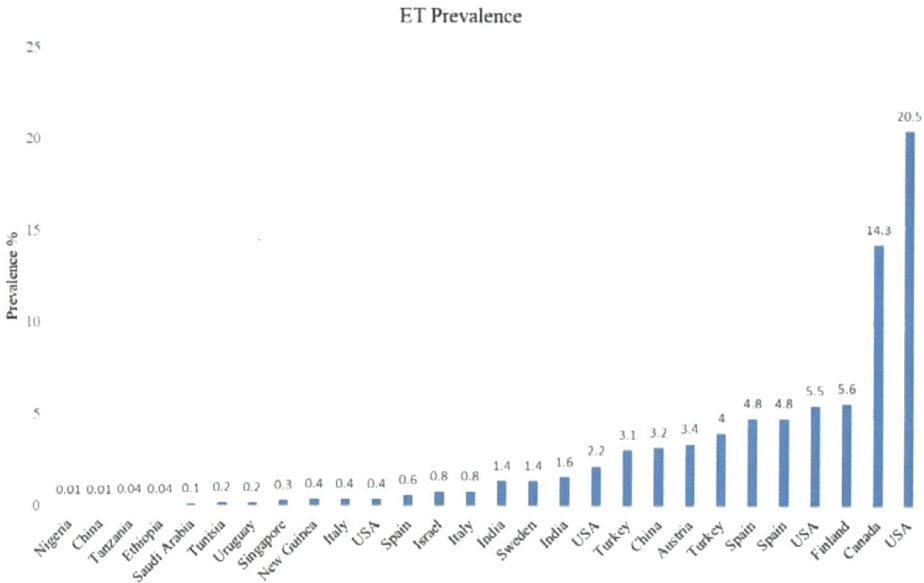

Fig. (1). Studies are ordered from lowest to highest prevalence (expressed in %). There are differences between studies regarding the screening process and epidemiological methods, in addition of variability in terms of age of examined subjects (notably, Canadian and US studies gather mainly patients in the elderly) [Adapted from [11]].

worldwide prevalence was then estimated at 0.9%. The prevalence markedly increased with age (4.6% for age ≥ 65 years), and especially with advanced age.

Although data are quite incomplete in numerous cases, the prevalence of ET may show regional and possibly ethnic differences. A majority of studies do not show a gender difference in ET prevalence (ratio men to women of 1.08:1.00); 6 studies indicate a gender imbalance (5 studies; ratio men: women 1.65:1.00) whereas one study indicates higher prevalence in women (ratio men: women 0.39:1.00) [11].

The very high incidence and prevalence raise the question of a predisposition to the disease [8]. Age is clearly a risk factor for ET. Most studies indicate a marked age-associated rise so that prevalence may be higher than 20% in the oldest patients [11].

PATHOGENESIS

ET is a complex disorder, poorly understood both in terms of etiology or pathophysiology. Epidemiological data suggest that a combination of genetic abnormalities with putative non-genetic (*e.g.*, environmental) factors lead to a slowly progressive neurodegeneration that causes shaking and other disturbances of neurologic function.

Many hypotheses relate to the etiology of ET, with two dominant central models – a conventional physiological model, also called *"olivary model"*, and the more recent *"degenerative cerebellar model"*, underpinned by molecular mechanism, cell biology and anatomo-pathology.

In the first model, ET would be in essence a primary electrical/electro physiological disorder, resulting from the overactivity of pacemaking neurons located in the inferior olivary nucleus. These neurons, which are coupled by gap junctions, fire in a rhythmic manner and target the Purkinje neurons in both the cerebellar cortex and cerebellar nuclei. The over-activity of the inferior olive would lead to a rhythmical burst firing in the cerebellar cortex and cerebellar nuclei (the sole output of the cerebellar circuitry), therefore producing tremor through an abnormal olivo–cerebellar activity and *via* the cerebello–thalamo–cortical output channels [12].

This model, which puts forward the inferior olive as the pacemaker of ET, was recently surpassed by a degenerative cerebellar cortical model based on intensive tissue-based studies that identified structural changes within the cerebellar cortex circuitry itself. Indeed, a loss of Purkinje cells was demonstrated by post-mortem investigations in ET cases using cells count, as well as linear density measurements [13]. The population of Purkinje cells would in fact represent the site of initial molecular/cellular events (hence the terminology of *"purkinjopathy"*) [14], generating a secondary remodeling/rewiring within the cerebellar cortex, with subsequent changes in adjacent neuronal populations (mainly the interneurons surrounding the Purkinje neurons). The formation of this aberrant cerebellar circuitry is probably central to the pathogenesis of ET, with notably thickened axons and remodeled basket cells [15, 16].

ET, as a progressive, age-related, disease, appears indeed truly *neurodegenerative* in nature [17]. This theory is further supported by evidence of brain iron accumulation [18]. Such iron deposits have been observed in other neurodegenerative disorders, such as Alzheimer's, Parkinson's and Huntington's diseases, that are also progressive disorders associated with ageing. In all these devastating disorders, cell loss occurs in combination with other cerebral changes or deposits (such as Lewy bodies for instance) [19]. It is noteworthy that the phenotype of these diseases includes a constellation of motor/non-motor dysfunctions [20].

However, the neurodegenerative hypothesis does not explain the early onset cases and the clinically very slow and heterogeneous progression of ET in other cases. Some patients show combinations of tremor and minor cerebellar symptoms with no evidence of other brain structures involvement. Furthermore, although there are clinical and electrophysiological arguments of cerebellar manifestations in ET patients, the reversibility of these symptoms and signs by ethanol intake (or thalamic deep brain stimulation: DBS; this is a matter of debate) challenges the neurodegenerative hypothesis [21, 22]. Some may claim that DBS is very active in Parkinson's disease whose pathogenesis is clearly neurodegenerative. Moreover, it can be argued that the Purkinje cell loss could be the result of long-standing tremor and not its cause (given that the cerebellum receives numerous afferences *via* the spinocerebellar pathways and the pontocerebellar tracts), although one would expect a progressive cerebellar degeneration in all steady tremor conditions if this were true [5].

An alternative hypothesis to ET genesis is based on inherent neural instability that leads to the generation of rhythmic bursts in central oscillating pacemakers such as thalamic nuclei (neurons of the inferior olivary complex also fall in this category). Through their tight interconnections within motor system networks, these oscillators become entrained, coupling their firing patterns to result in visible and pathologic tremors. This view can account at least partially for the heterogeneity of ET manifestations and therapeutic responses. Although structural alterations are not a pre-requisite for a neural instability, a pre-existing structural damage in the brain is a likely ground [5, 23]. Reports highlight that key structures involved in tremor genesis such as thalamic nuclei and the inferior olivary complex are normal. Furthermore, the fact that degeneration of inferior olive does not lessen tremor reinforces the concept that the inferior olivary nucleus does not play a critical role in the generation of tremor in these patients [24].

Neuroimaging studies have provided arguments for a disorder of the CNS. Reported outcomes from fMRI (functional Magnetic Resonance Imaging) and cortico-muscular MEG (Magneto Encephalography) analyses indicate that ET is mainly a disease of central origin, involving in particular 3 key-nodes within the numerous brain networks: the cerebellum, the thalamus and the primary motor cortex [25, 26]. PET (Positron Emission Tomography) studies suggest a gamma-aminobutyric acid (GABA)ergic dysfunction in tremor generation. A correlation has been identified between flumazenil uptake (flumazenil is a selective benzodiazepine receptor antagonist) and tremor rating scales, pointing towards abnormalities in GABA receptor binding. This defect would lead to a lack of inhibition within the cerebellar microcircuits, especially at level of the cortico-nuclear synapses between Purkinje and cerebellar nuclei neurons; the resulting glutamatergic overactivity (disinhibition of cerebellar nuclei) would generate tremor along the cerebello-thalamo-cortical pathways [27, 28]. This mechanism is supported by the observation that ET is highly responsive to ethanol, benzodia-zepines and barbiturates, which all facilitate inhibitory neurotransmission by binding to the $GABA_A$ receptor in the brain [29]. A pharmacological correction of GABA dysfunction could thus have a potential therapeutic effect in ET.

Etiology

ET might be triggered by a combination of both intrinsic and extrinsic mechanisms. Regarding the latter, environmental risk factors probably contribute to the etiology in a considerable proportion of cases. Yet there has been relatively little discussion in the tremor literature about a clear identification of these factors. Exposure to beta-carbolines, mercury, lead, organochlorine pesticides have all been incriminated [30]. Harmane, a heterocyclic amine (HCA) β-carboline alkaloid (βCA), is a potent tremor-producing neurotoxin. It is often found in the human diet and therefore a lifelong exposure is plausible. Blood concentrations are elevated in patients with ET as compared with controls and increased blood harmane concentration could be associated with cerebellar neuronal damage [31]. Louis *et al.* have demonstrated a strong inverse correlation between cerebellar N-acetyl aspartate to creatine ratio (NAA/Cr; a marker of neuronal damage) and blood harmane concentrations in 12 ET cases [31]. The correlation was absent in other brain regions such as thalamus and basal ganglia, or with other neurotoxins such as lead or manganese. In addition, animal studies have demonstrated that harmane and other βCAs produce cerebral damage. However, further confirmations on human post-mortem tissues, including accurate measurements of βCAs levels, are needed [30]. In order to avoid artifactual βCAs formation, tissues from autopsies should be preserved without formol/formaline solutions.

EXPERIMENTAL ANIMAL MODELS OF ET

Animal models of tremor have been developed in experimental neurology, because they remain a cornerstone, not only for understanding the pathophysiology of human tremor disorders, but also for the development of novel therapeutic agents. At least two approaches have been used to trigger, in animals, tremor reminiscent of ET: (1) administration of tremorgenic drugs such as harmaline, (2) use of various inbred strains.

Harmaline-induced tremor in rodent has been proposed as a possible model of essential tremor [32]. There are similarities between the two conditions, in particular the attenuation induced by ethanol [33]. Harmaline induces an action tremor with both kinetic and postural components. Anatomically, neurons of the inferior olivary nucleus (ION) have excitatory projections to the Purkinje cells of the cerebellar cortex (climbing fibers). As mentioned earlier, the ION neurons are electrically coupled and generate synchronous oscillations of membrane potentials. Harmaline acts directly on ION neurons, modulating their rhythm-generating ionic currents and facilitating rhythmic discharges. In rodents, it is presumed that harmaline-induced bursting is transmitted from the cerebellum to motor neurons in the spinal cord *via* the brainstem, thus resulting in generalized tremor (Fig. **2**). However, three points have questioned the relevance of this model for the pathophysiology of ET, *(i)* the primary target of harmaline: the role of the ION neurons remains controversial in ET and harmaline interferes also directly with several brain neurotransmitters; *(ii)* the transmission pathways: in ET, the cerebello-thalamo-cortical pathways are considered as the main route of electrical bursts spreading from the cerebellar circuitry towards the motor cortex and subsequently from the motor cortex to the motor neurons of the spinal cord; and *(iii)* species-specific differences have been observed in the response of the olivocerebellar system to harmaline and in the vulnerability of the Purkinje cell layer [34] (Fig. **2**).

A survey of animal models with chronic partial purkinje cells loss has been reviewed [35] since clinical studies suggest relation between purkinje cells loss and ET [13]. There is a limitation in the reviewed studies as tremor was not the primary interest; in addition, there were no constant results. Some models with chronic severe loss of purkinje cells show no tremor *i.e.* Purkinje cell degeneration mouse where the lost purkinje cells axon terminals on DCN neurons are replaced by astrocytic glial leaflets [36]. While other models with acute severe loss or chronic partial loss of purkinje cells may display tremor, *i.e.* Weaver mouse, scrambler mice, sticky mouse, toppler mice, WDR81 mice, shaker rat, PC degeneration in cats [35]. For all those models, further studies are needed to determine the similarity of induced tremors with those observed in ET patients.

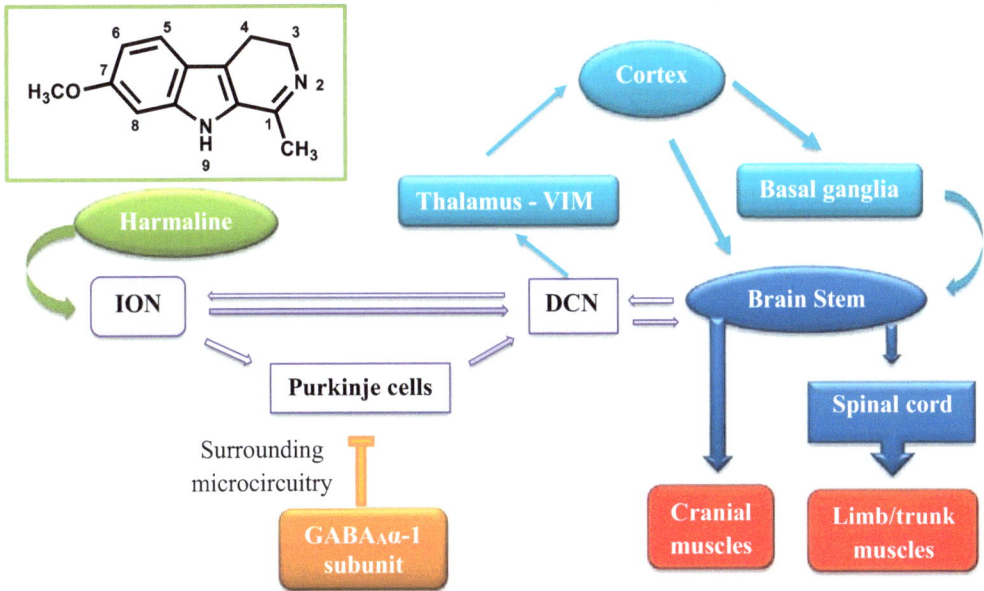

Fig. (2). Tremor-generating mechanisms and related structures in the CNS. Harmaline directly acts upon coupled neurons of the inferior olive (ION). Cytosolic pores are composed of the neuron-specific connexin 36 (Cx36) . Harmaline enhances neuronal synchrony and rhythmicity in the whole olivocerebellar system *via* the climbing fiber system. Deep cerebellar nuclei (DCN) project themselves back to the inferior olive *via* the inhibitory nucleo-olivary pathway. In GABA$_A$ receptor α-1 subunit knockout mice, neuronal response to synaptic GABA is lost in cerebellar Purkinje cells, resulting in rhythmical activities. VIM (Ventral intermediate nucleus).

The GABA$_A$ receptor α-1 subunit knockout mouse model [37] represents another rodent model of tremor, providing additional insight into the GABAergic mechanism involved in tremor genesis. Deletion of the GABA$_A$ receptor α-1 subunit produces a tremor with postural and kinetic components similar to essential tremor. In these mice, the response to synaptic and exogenous GABA is lost in cerebellar Purkinje cells, but the brain remains morphologically intact. As in the harmaline model, the tremor can be inhibited by ethanol consumption. This tremor is genetic and persistent, an advantage, compared to short-lived chemical-induced tremors. Moreover, the efficacy of the few drugs used in the treatment of human ET is also observed in α-1 knockout mice, lending further support to the model and possibly providing insight on ET-associated GABAergic dysfunction. However, this model should not be regarded as a genuine model of ET [38]. First, it has been shown that genetic mutations in the GABA$_A$ receptor α-1 gene have likely no significance in ET. Indeed, the frequencies of the GABRR1 (GABA receptor subtype rho1) genotypes and allelic variants do not differ between ET patients and control subjects [39]. Secondly, the onset of ET generally appears in elderly population and just occasionally during childhood, while tremor occurs early in development in these knockout mice. Finally, there are noticeable differences regarding the tremor frequency (knockout mouse: 16–22 Hz; ET: 4–12 Hz) [38]. Tremor frequency is known to be related in particular to the biomechanical features of limbs, such as inertia which is much higher in human.

As none of the animal models completely mimics the phenotype of human ET or recapitulates its histopathology, this clearly limits the prospects of discovering effective therapeutic agents. Also, to better understand the roles of the environment, new models are definitely needed. For instance, simple *in vitro* models, notably mini-brains, 3D lab-grown bundles of human brain cells that mimic the architecture of the cerebral cortex [40], could be adequate to assess neurotoxic agents and potential mechanisms leading to ET. This novel knowledge might help in devising new therapeutic options to identify neuroprotective measures for early-stage patients.

ET TREATMENT

There is still no cure for ET and no therapy has shown an effect on the reduction of the natural progression of the disease. Current symptomatic treatments aim to reduce the involuntary movements as much as possible, providing relief and improving the quality of life. Current therapies are based on drugs and surgical procedures. Therapeutical options are selected according to the severity of tremor and side effects.

Pharmacotherapy

All medications used to reduce tremor have initially been developed and approved for other indications. Unfortunately, the symptomatic drug benefit declines with time in all cases [41].

The first line therapy relies on propranolol and primidone [42]. Although propranolol is a well-known nonselective β-adrenergic receptor antagonist, the specific mechanism of its antitremor action has not been fully uncovered. The beneficial effect appears to be mainly due to blockade of peripheral beta-2 receptors on extrafusal muscle fibers and muscle spindles [43], although there may also be a synergistic CNS effect [44]. The daily dose varies from 60 to 800 mg/day with an average dose of 182.5 mg/day [45, 46]. There is no convincing evidence that doses higher than 320 mg/day may provide any additional benefit [46]. The proportion of responders varies from 50 to 70%, and the average tremor reduction is about 50% when compared with placebo [47]. Efficacy of both conventional and long acting propranolol is established only for tremor affecting the upper extremities, while the head tremor response is quite limited [45, 48]. Side effects include worsening of a pre-existing asthma, sinus bradycardia and fatigue. The β adrenergic antagonists atenolol (β-1 selective) and sotalol (nonselective) are also used for tremor control; atenolol is proposed for patients with an increased risk of bronchospasms [49].

The antitremorogenic action of primidone, an anticonvulsant of the barbiturate class which is metabolized into the active metabolites phenobarbital and phenylethylmalonamide (PEMA), is not fully understood either. Primidone reduces high-frequency repetitive firing of neurons and modifies transmembrane sodium and calcium channels ion movements, a possible mechanism for both its anticonvulsant and antitremor activities [50]. The daily doses range from 50 to 1000 mg/day and the average dose is around 500 mg/day. Average tremor improvement is up to 75% reduction from the baseline, even though most studies reported approximately 50% improvement when compared with placebo [45, 51]. Primidone presents a high incidence of adverse effects, such as nausea, ataxia and confusion, ranging from 22 to 72% of patients, resulting in a dropout rate from therapeutic studies ranging from 20 to 30% [52, 53]. Primidone and propranolol may be used in combination to treat limb tremor when monotherapy does not sufficiently reduce tremor [42].

The second line therapy for ET includes alprazolam, gabapentin, topiramate and clozapine [42]. The benzodiazepine alprazolam is an allosteric modulator of the GABAergic neurotransmission, potentiating the influx of chloride ions. The resulting hyperpolarization of the cell membrane inhibits action potential firing [54]. Alprazolam at 0.125 to 3 mg/day reduces limb tremor intensity by 25 to 34%. Side effects are mild, with sedation and fatigue most common, reported in 50% of patients [55]. The risk of drug abuse should not be underestimated. Gabapentin, an anticonvulsant with a structure similar to GABA, probably interacts with auxiliary subunits of voltage-gated calcium channels [56]. According to some studies, gabapentin reduces tremor intensity by 77% when used as monotherapy in doses of 1,200 mg/day [57]. Topiramate presents complex mechanisms of anticonvulsant action, but it remains unknown which mechanism plays a role in tremor control. Its use is limited by the high incidence of adverse effects [58].Clozapine, an atypical neuroleptic, is recommended only for refractory cases of limb tremor in ET due to the rare but serious risk of agranulocytosis [42, 59].

Ethanol decreases tremor severity in up to 50% of ET patients [60], but the side effects of sedation and intoxication clearly limit its use. However, this has led to new trials for the treatment of ET, especially with long-chain alcohols such as 1-octanol and its metabolite octanoic acid (OA). 1-octanol was demonstrated to be safe and effective with excellent tolerability but the large volumes to be administered when formulated in capsules seem to limit further development as an effective treatment [61, 62]. Preclinical and early-stage clinical trial data indicate a promising efficacy and acceptable safety for OA, the 1-octanol active metabolite, with more favorable pharmacological properties for drug delivery. However, further studies on long-term safety and efficacy of OA are still needed [63, 64].

A future therapeutic option could be based on the administration of vanillin, a commonly used food additive and flavoring agent. Experimentally, vanillin reduces harmaline-induce tremors in rats. However, the mechanism of action remains unclear. A potent inhibitory effect on serotonin pathways in the brain has been suggested [65]. Trials in human are missing.

Surgical Treatment

Thalamotomy causes a lesion in the ventral intermediate nucleus (VIM) of the thalamus. The target area is stereotactically localized. Micro-electrodes recordings are used to identify the typical pattern of discharges, confirming the location of the target. Neurostimulation with a macroelectrode can be applied in the awake patient during surgery to estimate tremor reduction and side effects. Unilateral thalamotomy reduces contralateral limb tremor in 80 to 90% of patients with ET [66, 67]. Bilateral thalamotomy is rarely performed nowadays because of common and often severe side effects [68, 69].

Deep brain stimulation of the thalamus (DBS) uses high frequency electrical stimulation exerted *via* an implanted electrode in order to modify the activity of the target area. The exact mechanisms by which DBS suppresses tremor are unknown, and postmortem examinations have not shown any permanent anatomic changes other than the electrode tract [70]. In most cases four electrodes, placed in VIM at a distance of 1.5 mm from each other, are connected to a pulse generator implanted in the chest wall. Electrode montage, voltage, pulse frequency and pulse width can be adjusted to optimize tremor control [71]. This flexibility in placing and adjusting the "functional lesion" is the main advantage of DBS as compared to thalamotomy which causes an irreversible lesion. Potential disadvantages of DBS include the higher cost and effort in programming and maintaining the device, in addition to dysfunction of the device related to cables. Following unilateral and bilateral DBS, mean tremor improvement reaches up to 60 to 90% on clinical rating scales as compared to baseline [72, 73].

Both techniques are invasive neurosurgical procedures. A long-term study of thalamotomy and DBS indicates that, although the benefits continue in most patients, a certain percentage of patients show a decline in response over time. This percentage of tremor recurrence has been reported as high as 35% in DBS [74]. The tremor recurrence can sometimes be effectively treated by changing the parameters of stimulation in patients who have undergone DBS but not in case of thalamotomy. A comparative study on 68 patients concludes that thalamic stimulation and thalamotomy are equally effective for the suppression of drug-resistant tremor, but thalamic stimulation has fewer adverse effects and results in a greater improvement of ET symptoms [66].

Gamma knife surgery (GK) is a non-invasive treatment based on radiation beams, from multiple angles, to an intracranial target based on anatomical imaging. In the case of ET, the target is the VIM. Alone, each beam is too weak to damage the healthy tissue through which it travels. However, the combined radiation is strong enough at the crossings of the beams to generate a local lesion. Several studies have found favorable results with gamma knife thalamotomy but the clinical improvement can take weeks to months. In follow-up studies, 92.1% of patients were entirely or nearly tremor-free postoperatively, and 88.2% remained tremor-free four years after the GK [75, 76]. Unfortunately delayed complications, such as complex movement disorders, have been reported [77].

Repetitive transcranial magnetic stimulation (rTMS) utilizes an electromagnet placed on the scalp to generate magnetic field pulses possessing roughly the strength of an MRI scan. The magnetic pulses stimulate an area of about 2.5 cm diameter on the surface of the brain. At low frequency (1 Hz), TMS induces small, sustained reductions in activity in the stimulated part of the brain. Low-frequency rTMS of the cerebellum can effectively modulate the cerebellar output, significantly improving total and specific (tremor, drawing, functional disability) scores, and reducing tremor amplitude [78]. However, a large clinical trial is still missing.

Another evolving technique of cerebellar neurostimulator is transcranial direct current stimulation (tDCS) [79], a powerful tool for the modulation of the cerebellar cortex excitability. The current (usually between 1 and 2.5 mAmp) is delivered at various sites, including the cerebellum and the frontal lobe [80]. Results obtained with the technique of transcranial alternating current stimulation (tACS) suggest that a single neural oscillator insures the temporal stability of ET tremor *versus* parkinsonian tremor frequency. There is a genuine hope that these techniques will be refined in the next years to reduce ET [81].

BETA-CARBOLINES AND THEIR PUTATIVE MECHANISMS OF ACTION

The βCAs are a group of indole alkaloids that notably includes harmane, harmine and harmaline (Fig. **3**). βCAs exhibit potent biological, psychopharmacological and toxicological activities. They occur naturally in plants, foods, and can be formed endogenously in mammals and humans [82].

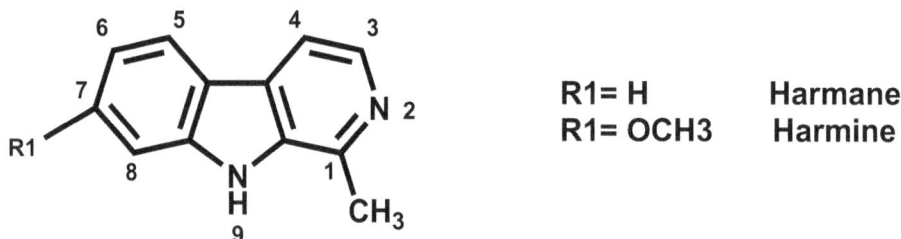

R1= H   Harmane
R1= OCH3   Harmine

Fig. (3). Chemical structures of major β-carboline alkaloids (βCAs).

Structurally, βCAs are heterocyclic amines, consisting of a combination of five- and six-ringed cycles, containing 2 amine groups. There is some structural similarity with 1-methyl-4-phenyl-1,2,3,6-tetrahydropyridine (MPTP), which is commonly used to produce major toxin-induced animal models for Parkinson's disease [83]. Like MPTP, the βCAs are highly neurotoxic.

Harmane crosses the blood-brain barrier, through an active uptake mechanism, and concentrates in the brain [84]. Laboratory animals exposed to harmane and other heterocyclic amines develop an intense and generalized action tremor a few minutes after administration. Tremor resembles ET [85] and is accompanied by destruction of cerebellar tissues [86, 87]. The increased blood harmane concentrations in ET patients have clearly generated an interest in the pathogenesis. However, the mechanisms behind this observation are not clear. Increased chronic dietary intake and/or genetic-metabolic factors could be involved concomitantly [86, 88, 89]. Indeed, harmane is particularly abundant in meats, and its concentration is increased by certain cooking practices (*e.g.*, long cooking times, over-cooking) [82, 90], notably through the Maillard reaction, a succession of non-enzymatic glycation thermal reactions that provide the basis for the colors and aromas characteristic of cooked foods. This complex network of reactions, that begins by condensation of sugars with amino groups of proteins, peptides or amino acids, is followed by rearrangement into Amadori-/Heyns- and reductone-type products; these induce degradation of amino-acids to yield aldehydes that can condense with amines and cyclize through a Mannich-type reaction, eventually leading to the formation of various βCAs, including harmane (Fig. **4**) [91, 92].

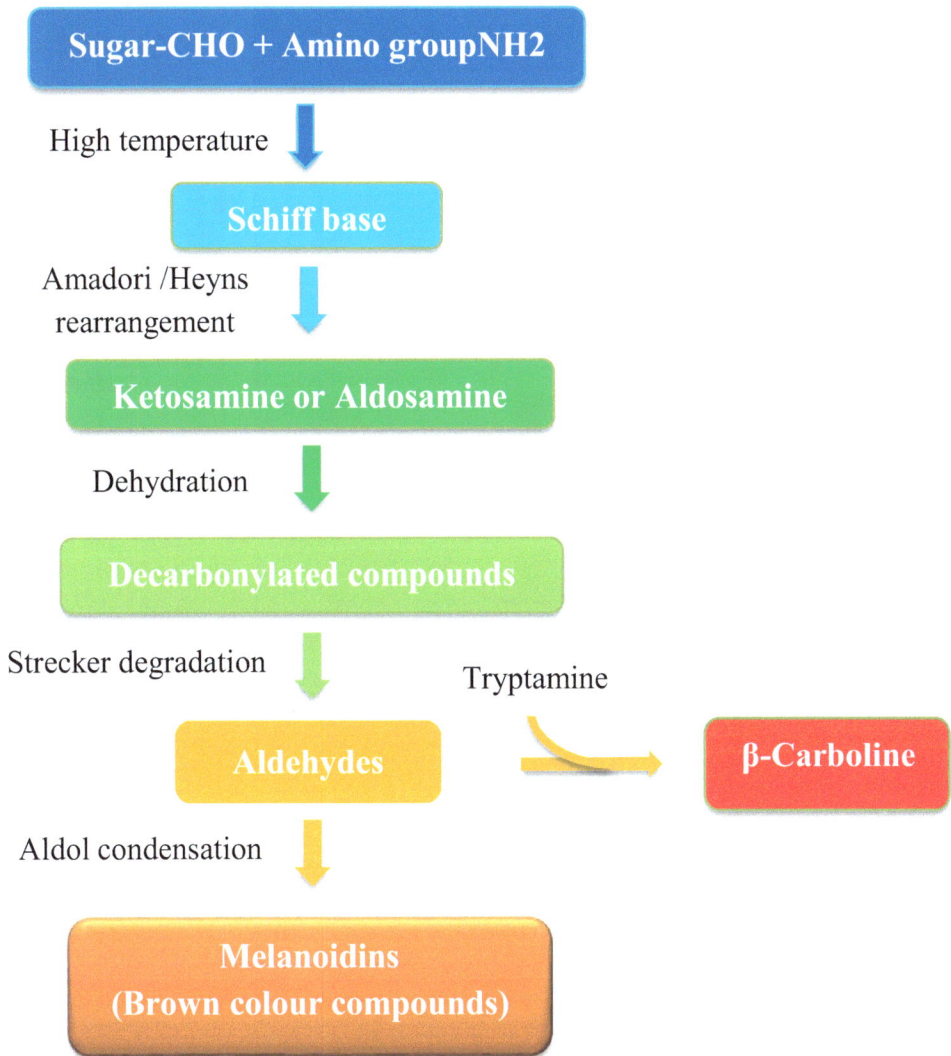

Fig. (4). Suggested pathway for the formation of β-carbolines through Maillard reaction.

This alkaloid can also be endogenously generated in human tissues and brain through a Mannich-type reaction arising from aldehydes condensation with tryptamine derivatives (Fig. **5**) [93].

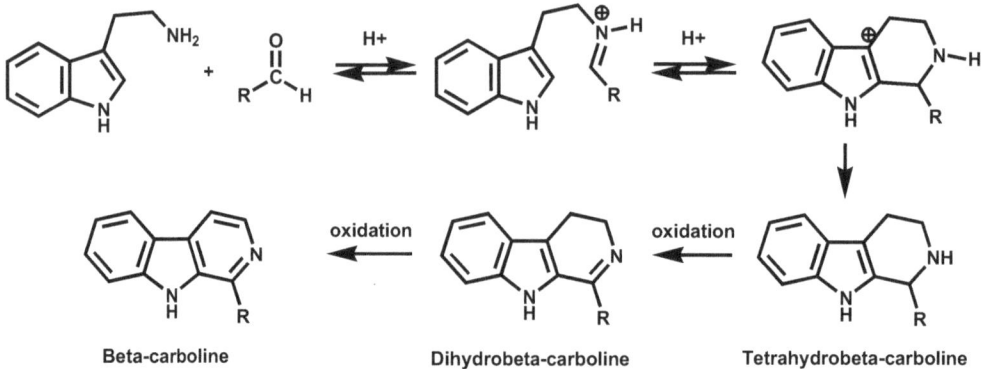

Fig. (5). Pathway for endogenous synthesis of βCAs by condensation of endogenous tryptamine with aldehydes or keto-acids. Similar reaction could occur with tryptamine derivatives, such as serotonin.

Bioactivation and Metabolization of Beta-Carbolines

MPTP Fig. (6) is well known to be neurotoxic. Its bioactivation to MPP$^+$ is carried out by the MAO_B and MPP$^+$ is selectively absorbed into the nigrostriatal dopaminergic neurons *via* a dopamine transporter (DAT). Given some structural analogy with MPTP, the potential neurotoxic effects of βCAs were investigated [94]. βCAs are usually less toxic than MPTP/MPP+. As pointed out above, βCAs occur naturally ", have been detected in human brain and might contribute to the degeneration of dopaminergic neurons during chronic exposure [95]. βCAs are also transported by DAT, but with lower efficiency than dopamine. They present enhanced cytotoxicity in DAT-expressing cells. Nevertheless, the low affinities of βCAs to DAT suggest other absorption pathways of neurotoxic βCA$^+$s, independent of DAT [96].

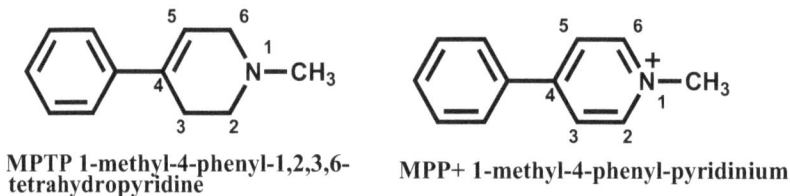

MPTP 1-methyl-4-phenyl-1,2,3,6-tetrahydropyridine   MPP+ 1-methyl-4-phenyl-pyridinium

Fig. (6). Chemical structure of MPTP and MPP$^+$.

The neurotoxicity of endogenous and exogenous βCAs is then probably affected by several factors, including their bioavailability, their toxic potential, their bioactivation/metabolism and their affinity for DAT.

The bioactivation of βCAs by N-methyltransferases (NMT) into the cationic neurotoxicants 2-ME-βCA$^+$s and 2,9-diME-βCA$^+$s is important for their relocalization to the brain and neurotoxicity [97]. Although the N-methylation may occur on both nitrogens, the methylation of the indole nitrogen appears to be the rate-limiting step in the development of toxicity [98]. Toxicity increases for βCAs methoxylated on the indole ring [95, 97].

Oxidative metabolism is another route for beta-carboline bioactivation [93, 99], through a reaction catalyzed by heme peroxidases, including myeloperoxidase and lactoperoxidase; these could be key catalysts for the bioactivation of endogenous and naturally occurring N-methyl-tetrahydro-βCAs [93, 99].

Metabolisation of βCAs in the liver and peripheral tissues by P450 enzymes (CYP1A, CYP1A2, CYP2C9, CYP2C19 and/or CYP2D6, depending on the βCA) may serve as detoxification routes, leading to hydroxylated βCAs, suggesting a possible role for cytochromes in protecting from this neurotoxin [100].

Beta-Carbolines: Neurotoxin or Neuroprotective?

Neurotoxicity of βCAs depends on the dose. High or chronic doses trigger neurotoxicity [98]. By contrast, low doses may increase dopamine levels and perhaps even present protective properties [101].

To achieve neuroprotection, both enzymes MAO_A and MAO_B should be inhibited to a certain level. This inhibition decreases the production of detrimental reactive oxygen species (ROS), a primary factor in neurodegeneration. βCAs can inhibit MAOs. Interestingly, βCAs (norharman and harman) have been identified in cigarette smoke [102]. This inhibition may explain the reduced risk of Parkinson's disease observed in smokers [103]. Nevertheless, it should be emphasized that, although neurodegenerative diseases share many pathological features like oxidative stress, iron accumulation, excitotoxicity and elevated ROS production [104], the neuroprotective action of tobacco smoke cannot be generalized to other neurodegenerative diseases, for example Alzheimer's disease [105].

An example of protective effects is the administration of harmine to a rat model of global cerebral ischemia. Harmine attenuates cerebral infarct volume and decreases neuronal death. It also causes a significant elevation of the glutamate transporter-1 (GLT-1) mRNA/protein and a remarkable attenuation of astrocyte activation [106]. In addition, harmine induces up-regulation of GLT-1, a neuroprotective effects in a rat model of amyotrophic lateral sclerosis disease [107]. GLT-1 dysfunction has been shown in the pathogenesis of multiple neurological disorders, including stroke and Alzheimer's disease. These findings certainly warrant further studies.

Moreover, calcium has been incriminated in the pathogenesis of neuro-degeneration. Since GABA pathways are involved in the control of calcium influx, directly *via* GABAergic receptors and indirectly *via* astrocytes and glial networks [108], the modulation of GABA transmission is potentially interesting when a neuroprotection is envisioned. It would be worth studying the putative relationship with βCAs, since the hypothesis of an intra-cellular calcium imbalance may apply to all human degenerative processes [109]. Also, the relationships between βCAs and L-glutamic acid decarboxylases isoforms GAD67 and GAD65 (molecular weights of 67 and 65 kDa, respectively) should be further investigated, since these isoforms are responsible for regulating the biosynthesis of GABA and its packaging into synaptic vesicles [110]. Very interestingly, GAD inhibition triggers a NMDA-mediated neuronal degeneration [111]. Recent reports highlight that MDMA (3,4-methylene-di-oxy- metham-phetamine) reduces GAD67 in the hippocampus, with an increase in seizure susceptibility involving glutamate receptor activation [112]. This confirms the importance of GAD67 in the homeostasy of the balance GABA/glutamate. Harmaline increases the extra-cellular concentrations of glutamate in cerebellar nuclei and impairs the NMDA-mediated regulation of glutamate (Fig. 7) [113]. As discussed here over, ET is highly responsive to alcohol, benzodiazepines and barbiturates, which all facilitate inhibitory neurotransmission by binding to the $GABA_A$ receptor in the brain [29]; a pharmacological correction of GABA dysfunction thus has a potential for ET therapy.

NEUROPROTECTIVE PROPERTIES OF DIETARY PHYTO-CHEMICALS AND HERBAL THERAPIES IN ET?

A Novel and Under Recognized Avenue for Research

The search for effective treatments of ET is on-going; given the high worldwide prevalence of the disease, it is conceivable that some plants used in traditional medicines may be effective and yield clues to developing new pharmacotherapies. In neurodegenerative disorders, in addition to the importance of neurotrophins levels, the role of oxidative stress and inflammation factors has been proven [114]. Both dietary phytochemicals (such as phenolic acids, catechins, flavonoids, anthocyans, stilbenoids, curcuminoids) and herbal therapies have a strong potential neuroprotective effect. From the extensive list of herbal medicines proposed to treat neurological diseases [115], some are noteworthy; but, most often, their mechanisms of action have not been determined and clinical studies are lacking.

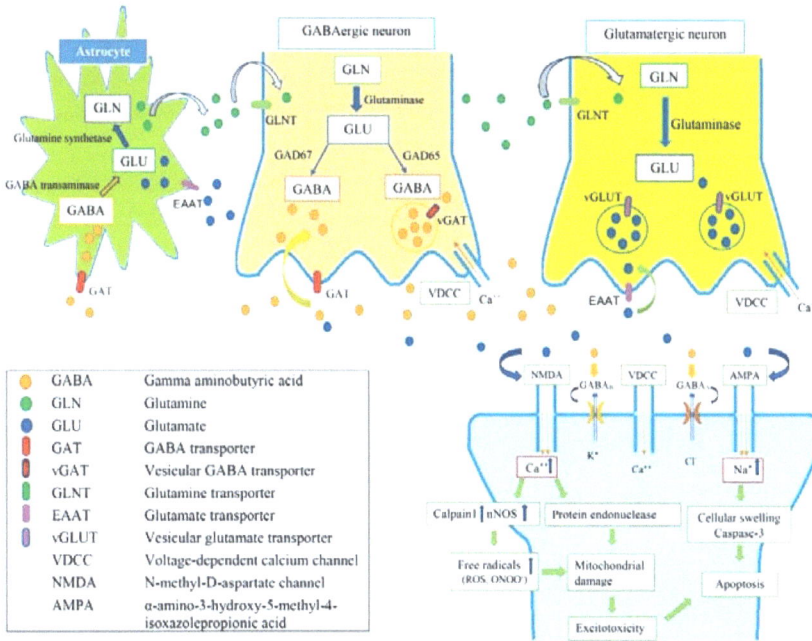

Fig. (7). Illustration of an astrocyte (green), a pre-synaptic Gabaergic (brown) neuron, a pre-synaptic Glutamatergic neuron (yellow) and a post-synaptic neuron (blue). Overstimulation of glutamatergic receptors results in an excitotoxic cascade triggered by an excess of calcium entry at the post-synaptic site.

Plants Well-Known for Neuroprotective Effects

Wild Syrian rue (*Peganum harmala* L.) seeds, bark, and root have demonstrated different pharmacological and therapeutic effects in the fields of cardiovascular, gastrointestinal, endocrine, cancer, pain, diabetes, respiratory, antimicrobial, anti-inflammatory, febrifuge and, mainly, nervous diseases [116]. The most important phytochemicals of the different parts of the plant are beta-carboline..... [117], especially harmalol, harmaline and harmine, and quinazoline [118] alkaloids. As discussed in the previous section, it is debated whether βCAs are neurotoxic or neuroprotective; this probably depends on their structural features, dose and duration of use. βCAs are inhibitors of MAO and interact with CNS receptors of opioids, dopamine, GABA, 5-hydroxytryptamine, and benzodiazepines, which may explain their numerous pharmacological effects [116].

The whole plant of *Bacopa monnieri* (L.) Wettst., traditionally used in India for longevity and cognitive enhancement is certainly one of the most promising herbal medicines in the neurology field. Neuroprotective effects of its extracts include antioxidant/neuroprotection, acetylcholinesterase inhibition, choline acetyltransferase activation, β-amyloid reduction, increased cerebral blood flow, and monoamine potentiation and modulation. The active constituents, bacosides, triterpenoid saponins with at least twelve known analogs [119], are heavily metabolized; a major metabolite, ebelin lactone could be the main neuroprotective agent [120]

The leaves of *Ginkgo biloba* L. harbor two major classes of active components, flavonoids and ginkgolides, known for their neuroprotective properties through different possible mechanism of action, PAF antagonism, ROS and NO scavenging, interaction with neurotransmitters and induction of growth factors [121, 122].

In the Ayurveda remedy, *mandukaparni* (*Centella asiatica* (L.) Urb.), the major constituents are saponins (medacoside, asiaticoside, medacassoside, based on asiatic acid) and polyphenols [123]. The chloroform: methanolic extract showed neuroprotective efficacy due to free radical scavenging by the polyphenols and flavonoids [124].

Rubia cordifolia L., *Fagonia cretica* L. and *Tinospora cordifolia* (Willd.) Miers (a synonym of *Tinospora sinensis* (Lour.) Merr.) have been reported to be a rich source of antioxidants [125 - 127]. They act by reducing oxidant levels through direct scavenging to protect cells from free radicals generated during immune activation, which explains their properties as antioxidant and anti-inflammatory [128].

Phytochemicals Reported for Neuroprotective Effects

Many phytochemicals have been reported to exert neuroprotective effects [129], modulating some changes observed in neurodegenerative disorders.

Dietary polyphenolic compounds (phenolic acids, flavonoids,...) have shown a high efficacy as antioxidant and neuroprotectors [130] and many studies have reviewed their efficacy against neuroinflammation and apoptosis [131]. For example, grape polyphenols modified by fermentation, as found in red wine, are protective against neuronal death induced by 3-morpholinosydnonimine (SIN-1) in a dopaminergic cell line [132].

Studies indicate that, in addition to their antioxidant activity, catechins, the phenolic compounds of green tea, are capable to modulate the signal transduction pathways that can exert cell-survival and anti-inflammatory actions [133].

As neuroprotective agent, however, polyphenols such as flavonoids must cross the brain blood barrier (BBB) to protect vulnerable neurons [134]. Experimental data suggest that intact anthocyanins and/or their metabolites do enter the brain of rats [135] and pigs [136] fed diets supplemented with anthocyanins-rich extracts; in cellular models, these were found to enhance the mitochondrial function, a possible mechanism for neuroprotection [137].

Stilbenoids consist in a family of resveratrol derivatives. Resveratrol modulates multiple mechanisms important in neurodegenerative diseases, notably protecting dopaminergic neurons against metabolic and oxidative insults, in models relevant to Parkinson's and Alzheimer's diseases [138].

Curcumin has shown potent antioxidant, anti–inflammatory and anti–protein-aggregate activities [139] but poor BBB penetration. Nevertheless, many studies have investigated the role of curcumin in neurodegenerative disorders, indicating that curcumin may stop death cascades in neurodegenerative disorders, preventing the death of cells from inflammation and oxidative stress and reducing the aggregation of α-synuclein, a major component of the Lewy body lesions [140]. Various curcuminoids are being investigated for enhanced BBB penetration [141].

Given the prevailing hypothesis of a slow progressive neurodegenerative mechanism in ET, there is a definite need to explore whether a genuine neuroprotection would occur with dietary herbs and/or medicinal plants. A chronic administration in a cohort of patients could be envisioned, comparing with a matched control group.

CONCLUSION

ET is a highly prevalent neurological disorder affecting in particular the elderly, impairing the quality of life and whose pathogenesis remains poorly understood. ET is clinically heterogeneous, combining motor features with various cognitive and psychiatric signs. Both genetic and environmental factors probably contribute to the mechanisms of this slowly progressive neurodegenerative disease. Oxidative stress and inflammation factors leading to overactivation of glutamatergic pathways are suspected to explain the neurodegeneration process occurring in the brain, especially at the level of Purkinje neurons in the cerebellar cortex. Several studies point out the probable implication of dietary βCAs, formed in some overcooked foods especially *via* the Maillard reaction, in the cascade of events leading to tremor. The current lack of satisfying animal models impacts on the development of effective therapeutic options and we still miss effective strategies of prevention. Surprisingly, none of the therapies currently administered (drugs, surgical procedures) in patients target a neuroprotective effect. Since both dietary phytochemicals and herbal therapies may have a strong potential neuroprotective action, we suggest devising studies testing the hypothesis that efficient herbs will yield important clues to develop novel pharmacotherapies. Furthermore, we have discussed the dysfunction of GABA pathways and we have stressed the importance of GAD enzymes. Those enzymes might be a candidate target to restore the GABA/Glutamate balance. The era of neuroprotection is now emerging for one of the commonest movement disorders.

ABBREVIATIONS

ADL	Activities of Daily Living
AMPA	α-amino-3-hydroxy-5-methyl-4-isoxazole propionic acid
BBB	Brain Blood Barrier
βCA	Beta Carboline Alkaloid
CNS	Central Nervous System
DAT	Dopamine Transporter
DBS	Deep Brain Stimulation
DCN	Deep Cerebellar Nuclei
ET	Essential Tremor
EAAT	Excitatory Amino Acid Transporter
fMRI	Functional Magnetic Resonance Imaging
GABA	Gamma Aminobutyric Acid
GABA$_A$	Gamma Aminobutyric Acid A
GABA$_B$	Gamma Aminobutyric Acid B
GABRR1	GABA Receptor Subtype Rho1
GAD	Glutamic Acid Decarboxylase
GAT	GABA Transporter
GK	Gamma Knife Surgery
GLN	Glutamine
GLNT	Glutamine Transporter
GLT	Glutamate Transporter
GLU	Glutamate
HCA	Heterocyclic Amine
ION	Inferior Olivary Nucleus
MAO$_A$	Monoamine Oxidase A
MAO$_B$	Monoamine Oxidase B
MDMA	3,4-Methylenedioxy-methamphetamine
MEG	Magneto Encephalography
MPTP	1-methyl-4-phenyl-1,2,3,6-tetrahydropyridine

MPP+	1-methyl-4-phenyl-pyridinium
MRI	Magnetic Resonance Imaging
mRNA	Messenger Ribonucleic Acid
N-MeTH-βC	N-methyltetrahydrobetacarboline
NMDA	N-methyl-D-Aspartate Channel
NMT	N-methyltransferases
NO	Nitric Oxide
OA	Octanoic Acid
PAF	Platelet-Activating Factor
PEMA	Phenylethylmalonamide
PET	Positron Emission Tomography
ROS	Reactive Oxygen Species
SIN-1	3-morpholinosydnonimine
rTMS	Repetitive Transcranial Magnetic Stimulation
tACS	Transcranial Alternating Current Stimulation
tDCS	Transcranial Direct Current Stimulation
VDCC	Voltage-Dependent Calcium Channel
vGAT	Vesicular GABA Transporter
vGLUT	Vesicular Glutamate Transporter
VIM	Ventral Intermediate Nucleus

CONSENT FOR PUBLICATION

Not applicable.

CONFLICT OF INTEREST

The authors confirm that they have no conflict of interest to declare for this publication.

ACKNOWLEDGEMENTS

MM is supported by the FNRS and the Fonds Erasme. RA has been awarded a grant from the Fonds pour la Recherche Médicale dans le Hainaut (FRMH).

REFERENCES

[1]   Jankovic J. Essential tremor: clinical characteristics. Neurology 2000; 54(11) (Suppl. 4): S21-5. [http://dx.doi.org/10.1212/WNL.54.11.21A] [PMID: 10854348]

[2]   Brennan KC, Jurewicz EC, Ford B, Pullman SL, Louis ED. Is essential tremor predominantly a kinetic or a postural tremor? A clinical and electrophysiological study. Mov Disord 2002; 17(2): 313-6.

[http://dx.doi.org/10.1002/mds.10003] [PMID: 11921117]

[3]   Louis ED. Diagnosis and Management of Tremor CONTINUUM: Lifelong Learning in Neurology 2016; 22(4, Movement Disorders): 1143-58.
[http://dx.doi.org/10.1212/CON.0000000000000346]

[4]   Gasparini M, Bonifati V, Fabrizio E, *et al.* Frontal lobe dysfunction in essential tremor: a preliminary study. J Neurol 2001; 248(5): 399-402. [journal article].
[http://dx.doi.org/10.1007/s004150170181] [PMID: 11437162]

[5]   Deuschl G, Elble R. Essential tremor--neurodegenerative or nondegenerative disease towards a working definition of ET. Mov Disord 2009; 24(14): 2033-41.
[http://dx.doi.org/10.1002/mds.22755] [PMID: 19750493]

[6]   Sengul Y, Sengul HS, Yucekaya SK, *et al.* Cognitive functions, fatigue, depression, anxiety, and sleep disturbances: assessment of nonmotor features in young patients with essential tremor. Acta Neurol Belg 2015; 115(3): 281-7. [journal article].
[http://dx.doi.org/10.1007/s13760-014-0396-6] [PMID: 25471376]

[7]   Louis ED, Huey ED, Gerbin M, Viner AS. Depressive traits in essential tremor: impact on disability, quality of life, and medication adherence. Eur J Neurol 2012; 19(10): 1349-54.
[http://dx.doi.org/10.1111/j.1468-1331.2012.03774.x] [PMID: 22642492]

[8]   Louis ED. Essential tremor and other forms of kinetic tremor mechanisms and emerging therapies in tremor disorders.. New York, NY: Springer New York 2013; pp. 167-201.
[http://dx.doi.org/10.1007/978-1-4614-4027-7_10]

[9]   Louis ED. Essential tremor. Handbook of Clinical Neurology. Elsevier 2011; pp. 433-48.

[10] Benito-León J, Bermejo-Pareja F, Louis ED. Incidence of essential tremor in three elderly populations of central Spain. Neurology 2005; 64(10): 1721-5.
[http://dx.doi.org/10.1212/01.WNL.0000161852.70374.01] [PMID: 15911798]

[11] Louis ED, Ferreira JJ. How common is the most common adult movement disorder? Update on the worldwide prevalence of essential tremor. Mov Disord 2010; 25(5): 534-41.
[http://dx.doi.org/10.1002/mds.22838] [PMID: 20175185]

[12] Louis ED. Essential tremor as a neuropsychiatric disorder. J Neurol Sci 2010; 289(1-2): 144-8.
[http://dx.doi.org/10.1016/j.jns.2009.08.029] [PMID: 19720384]

[13] Axelrad JE, Louis ED, Honig LS, *et al.* Reduced Purkinje cell number in essential tremor: a postmortem study. Arch Neurol 2008; 65(1): 101-7.
[http://dx.doi.org/10.1001/archneurol.2007.8] [PMID: 18195146]

[14] Grimaldi G, Manto M. Is essential tremor a Purkinjopathy? The role of the cerebellar cortex in its pathogenesis. Mov Disord 2013; 28(13): 1759-61.
[http://dx.doi.org/10.1002/mds.25645] [PMID: 24114851]

[15] Louis ED. From neurons to neuron neighborhoods: the rewiring of the cerebellar cortex in essential tremor. Cerebellum 2014; 13(4): 501-12.
[http://dx.doi.org/10.1007/s12311-013-0545-0] [PMID: 24435423]

[16] Louis ED. Re-thinking the biology of essential tremor: from models to morphology. Parkinsonism Relat Disord 2014; 20 (Suppl. 1): S88-93.
[http://dx.doi.org/10.1016/S1353-8020(13)70023-3] [PMID: 24262197]

[17] Bhalsing KS, Saini J, Pal PK. Understanding the pathophysiology of essential tremor through advanced neuroimaging: a review. J Neurol Sci 2013; 335(1-2): 9-13.
[http://dx.doi.org/10.1016/j.jns.2013.09.003] [PMID: 24060292]

[18] Novellino F, Cherubini A, Chiriaco C, *et al.* Brain iron deposition in essential tremor: a quantitative 3-Tesla magnetic resonance imaging study. Mov Disord 2013; 28(2): 196-200.
[http://dx.doi.org/10.1002/mds.25263] [PMID: 23238868]

[19] Louis ED, Faust PL, Vonsattel JP, *et al*. Neuropathological changes in essential tremor: 33 cases compared with 21 controls. Brain 2007; 130(Pt 12): 3297-307.
[http://dx.doi.org/10.1093/brain/awm266] [PMID: 18025031]

[20] Bonuccelli U. Essential tremor is a neurodegenerative disease. J Neural Transm (Vienna) 2012; 119(11): 1383-7.
[http://dx.doi.org/10.1007/s00702-012-0878-8] [PMID: 23011236]

[21] Kronenbuerger M, Tronnier VM, Gerwig M, *et al*. Thalamic deep brain stimulation improves eyeblink conditioning deficits in essential tremor. Exp Neurol 2008; 211(2): 387-96.
[http://dx.doi.org/10.1016/j.expneurol.2008.02.002] [PMID: 18394604]

[22] Klebe S, Stolze H, Grensing K, Volkmann J, Wenzelburger R, Deuschl G. Influence of alcohol on gait in patients with essential tremor. Neurology 2005; 65(1): 96-101.
[http://dx.doi.org/10.1212/01.wnl.0000167550.97413.1f] [PMID: 16009892]

[23] Kurtis MM. Essential tremor: is it a neurodegenerative disease? No. J Neural Transm (Vienna) 2012; 119(11): 1375-81.
[http://dx.doi.org/10.1007/s00702-012-0875-y] [PMID: 22899276]

[24] Louis ED, Diaz DT, Kuo SH, Gan SR, Cortes EP, Vonsattel JPG, *et al*. Inferior Olivary nucleus degeneration does not lessen tremor in essential tremor Cerebellum Ataxias 2018; 5(1): 018-0080.
[http://dx.doi.org/10.1186/s40673-018-0080-3]

[25] Schnitzler A, Münks C, Butz M, Timmermann L, Gross J. Synchronized brain network associated with essential tremor as revealed by magnetoencephalography. Mov Disord 2009; 24(11): 1629-35.
[http://dx.doi.org/10.1002/mds.22633] [PMID: 19514010]

[26] Contarino MF, Groot PF, van der Meer JN, *et al*. Is there a role for combined EMG-fMRI in exploring the pathophysiology of essential tremor and improving functional neurosurgery? PLoS One 2012; 7(10): e46234.
[http://dx.doi.org/10.1371/journal.pone.0046234] [PMID: 23049695]

[27] Boecker H, Weindl A, Brooks DJ, *et al*. GABAergic dysfunction in essential tremor: an 11C-flumazenil PET study. J Nucl Med 2010; 51(7): 1030-5.
[http://dx.doi.org/10.2967/jnumed.109.074120] [PMID: 20554735]

[28] Gironell A, Figueiras FP, Pagonabarraga J, *et al*. Gaba and serotonin molecular neuroimaging in essential tremor: a clinical correlation study. Parkinsonism Relat Disord 2012; 18(7): 876-80.
[http://dx.doi.org/10.1016/j.parkreldis.2012.04.024] [PMID: 22595620]

[29] Cross AJ, Misra A, Sandilands A, Taylor MJ, Green AR. Effect of chlormethiazole, dizocilpine and pentobarbital on harmaline-induced increase of cerebellar cyclic GMP and tremor. Psychopharmacology (Berl) 1993; 111(1): 96-8.
[http://dx.doi.org/10.1007/BF02257413] [PMID: 7870940]

[30] Louis ED. Environmental epidemiology of essential tremor. Neuroepidemiology 2008; 31(3): 139-49.
[http://dx.doi.org/10.1159/000151523] [PMID: 18716411]

[31] Louis ED, Zheng W, Mao X, Shungu DC. Blood harmane is correlated with cerebellar metabolism in essential tremor: a pilot study. Neurology 2007; 69(6): 515-20.
[http://dx.doi.org/10.1212/01.wnl.0000266663.27398.9f] [PMID: 17679670]

[32] Wilms H, Sievers J, Deuschl G. Animal models of tremor. Mov Disord 1999; 14(4): 557-71.
[http://dx.doi.org/10.1002/1531-8257(199907)14:4<557::AID-MDS1004>3.0.CO;2-G] [PMID: 10435492]

[33] Rappaport MS, Gentry RT, Schneider DR, Dole VP. Ethanol effects on harmaline-induced tremor and increase of cerebellar cyclic GMP. Life Sci 1984; 34(1): 49-56.
[http://dx.doi.org/10.1016/0024-3205(84)90329-1] [PMID: 6319933]

[34] Miwa H, Kubo T, Suzuki A, Kihira T, Kondo T. A species-specific difference in the effects of

harmaline on the rodent olivocerebellar system. Brain Res 2006; 1068(1): 94-101.
[http://dx.doi.org/10.1016/j.brainres.2005.11.036] [PMID: 16405928]

[35]   Handforth A. Linking Essential Tremor to the Cerebellum-Animal Model Evidence. Cerebellum 2016; 15(3): 285-98.
[http://dx.doi.org/10.1007/s12311-015-0750-0] [PMID: 26660708]

[36]   Roffler-Tarlov S, Beart PM, O'Gorman S, Sidman RL. Neurochemical and morphological consequences of axon terminal degeneration in cerebellar deep nuclei of mice with inherited Purkinje cell degeneration. Brain Res 1979; 168(1): 75-95.
[http://dx.doi.org/10.1016/0006-8993(79)90129-X] [PMID: 455087]

[37]   Kralic JE, Criswell HE, Osterman JL, *et al.* Genetic essential tremor in γ-aminobutyric acidA receptor α1 subunit knockout mice. J Clin Invest 2005; 115(3): 774-9.
[http://dx.doi.org/10.1172/JCI200523625] [PMID: 15765150]

[38]   Jankovic J, Noebels JL. Genetic mouse models of essential tremor: are they essential? J Clin Invest 2005; 115(3): 584-6.
[http://dx.doi.org/10.1172/JCI24544] [PMID: 15765140]

[39]   García-Martín E, Martínez C, Alonso-Navarro H, *et al.* Gamma-aminobutyric acid (GABA) receptor rho (GABRR) polymorphisms and risk for essential tremor. J Neurol 2011; 258(2): 203-11.
[http://dx.doi.org/10.1007/s00415-010-5708-z] [PMID: 20820800]

[40]   Sartore RC, Cardoso SC, Lages YV, *et al.* Trace elements during primordial plexiform network formation in human cerebral organoids. PeerJ 2017; 5(5): e2927.
[http://dx.doi.org/10.7717/peerj.2927] [PMID: 28194309]

[41]   Elble RJ. Tremor: clinical features, pathophysiology, and treatment. Neurol Clin 2009; 27(3): 679-695, v-vi.
[http://dx.doi.org/10.1016/j.ncl.2009.04.003] [PMID: 19555826]

[42]   Zesiewicz TA, Elble R, Louis ED, *et al.* Practice parameter: therapies for essential tremor: report of the Quality Standards Subcommittee of the American Academy of Neurology. Neurology 2005; 64(12): 2008-20.
[http://dx.doi.org/10.1212/01.WNL.0000163769.28552.CD] [PMID: 15972843]

[43]   Abila B, Wilson JF, Marshall RW, Richens A. The tremorolytic action of beta-adrenoceptor blockers in essential, physiological and isoprenaline-induced tremor is mediated by beta-adrenoceptors located in a deep peripheral compartment. Br J Clin Pharmacol 1985; 20(4): 369-76.
[http://dx.doi.org/10.1111/j.1365-2125.1985.tb05079.x] [PMID: 2866785]

[44]   Sweetman SC, Ed. Martindale: The complete drug reference. 36th Revised., London: Pharmaceutical Press 2009.

[45]   Cleeves L, Findley LJ. Propranolol and propranolol-LA in essential tremor: a double blind comparative study. J Neurol Neurosurg Psychiatry 1988; 51(3): 379-84.
[http://dx.doi.org/10.1136/jnnp.51.3.379] [PMID: 3283296]

[46]   Koller WC. Dose-response relationship of propranolol in the treatment of essential tremor. Arch Neurol 1986; 43(1): 42-3.
[http://dx.doi.org/10.1001/archneur.1986.00520010038018] [PMID: 3942513]

[47]   Koller WC. Long-acting propranolol in essential tremor. Neurology 1985; 35(1): 108-10.
[http://dx.doi.org/10.1212/WNL.35.1.108] [PMID: 3965982]

[48]   Tolosa ES, Loewenson RB. Essential tremor: treatment with propranolol. Neurology 1975; 25(11): 1041-4.
[http://dx.doi.org/10.1212/WNL.25.11.1041] [PMID: 1237822]

[49]   Leigh PN, Jefferson D, Twomey A, Marsden CD. Beta-adrenoreceptor mechanisms in essential tremor; a double-blind placebo controlled trial of metoprolol, sotalol and atenolol. J Neurol Neurosurg Psychiatry 1983; 46(8): 710-5.

[http://dx.doi.org/10.1136/jnnp.46.8.710] [PMID: 6310053]

[50]   Guan XM, Peroutka SJ. Basic mechanisms of action of drugs used in the treatment of essential tremor. Clin Neuropharmacol 1990; 13(3): 210-23.
[http://dx.doi.org/10.1097/00002826-199006000-00003] [PMID: 1972653]

[51]   Findley LJ, Cleeves L, Calzetti S. Primidone in essential tremor of the hands and head: a double blind controlled clinical study. J Neurol Neurosurg Psychiatry 1985; 48(9): 911-5.
[http://dx.doi.org/10.1136/jnnp.48.9.911] [PMID: 3900296]

[52]   O'Suilleabhain P, Dewey RB Jr. Randomized trial comparing primidone initiation schedules for treating essential tremor. Mov Disord 2002; 17(2): 382-6.
[http://dx.doi.org/10.1002/mds.10083] [PMID: 11921128]

[53]   Serrano-Dueñas M. Use of primidone in low doses (250 mg/day) *versus* high doses (750 mg/day) in the management of essential tremor. Double-blind comparative study with one-year follow-up. Parkinsonism Relat Disord 2003; 10(1): 29-33.
[http://dx.doi.org/10.1016/S1353-8020(03)00070-1] [PMID: 14499204]

[54]   Riss J, Cloyd J, Gates J, Collins S. Benzodiazepines in epilepsy: pharmacology and pharmacokinetics. Acta Neurol Scand 2008; 118(2): 69-86.
[http://dx.doi.org/10.1111/j.1600-0404.2008.01004.x] [PMID: 18384456]

[55]   Gunal DI, Afşar N, Bekiroglu N, Aktan S. New alternative agents in essential tremor therapy: double-blind placebo-controlled study of alprazolam and acetazolamide. Neurol Sci 2000; 21(5): 315-7.
[http://dx.doi.org/10.1007/s100720070069] [PMID: 11286044]

[56]   Taylor CP, Gee NS, Su TZ, *et al.* A summary of mechanistic hypotheses of gabapentin pharmacology. Epilepsy Res 1998; 29(3): 233-49.
[http://dx.doi.org/10.1016/S0920-1211(97)00084-3] [PMID: 9551785]

[57]   Gironell A, Kulisevsky J, Barbanoj M, López-Villegas D, Hernández G, Pascual-Sedano B. A randomized placebo-controlled comparative trial of gabapentin and propranolol in essential tremor. Arch Neurol 1999; 56(4): 475-80.
[http://dx.doi.org/10.1001/archneur.56.4.475] [PMID: 10199338]

[58]   Ondo WG, Jankovic J, Connor GS, *et al.* Topiramate in essential tremor: a double-blind, placebo-controlled trial. Neurology 2006; 66(5): 672-7.
[http://dx.doi.org/10.1212/01.wnl.0000200779.03748.0f] [PMID: 16436648]

[59]   Ceravolo R, Salvetti S, Piccini P, Lucetti C, Gambaccini G, Bonuccelli U. Acute and chronic effects of clozapine in essential tremor. Mov Disord 1999; 14(3): 468-72.
[http://dx.doi.org/10.1002/1531-8257(199905)14:3<468::AID-MDS1013>3.0.CO;2-M]   [PMID: 10348471]

[60]   Knudsen K, Lorenz D, Deuschl G. A clinical test for the alcohol sensitivity of essential tremor. Mov Disord 2011; 26(12): 2291-5.
[http://dx.doi.org/10.1002/mds.23846] [PMID: 22021159]

[61]   Bushara KO, Goldstein SR, Grimes GJ Jr, Burstein AH, Hallett M. Pilot trial of 1-octanol in essential tremor. Neurology 2004; 62(1): 122-4.
[http://dx.doi.org/10.1212/01.WNL.0000101722.95137.19] [PMID: 14718713]

[62]   Nahab FB, Wittevrongel L, Ippolito D, *et al.* An open-label, single-dose, crossover study of the pharmacokinetics and metabolism of two oral formulations of 1-octanol in patients with essential tremor. Neurotherapeutics 2011; 8(4): 753-62.
[http://dx.doi.org/10.1007/s13311-011-0045-1] [PMID: 21594724]

[63]   Haubenberger D, Nahab FB, Voller B, Hallett M. Treatment of essential tremor with long-chain alcohols: still experimental or ready for prime time? tremor and other hyperkinetic movements 2014; 4 tre-04-211-4673-2

[64]   Nahab FB, Handforth A, Brown T, *et al.* Octanoic acid suppresses harmaline-induced tremor in mouse

model of essential tremor. Neurotherapeutics 2012; 9(3): 635-8.
[http://dx.doi.org/10.1007/s13311-012-0121-1] [PMID: 22454323]

[65]   Abdulrahman AA, Faisal K, Meshref AA, Arshaduddin M. Low-dose acute vanillin is beneficial against harmaline-induced tremors in rats. Neurol Res 2017; 39(3): 264-70.
[http://dx.doi.org/10.1080/01616412.2016.1275456] [PMID: 28095756]

[66]   Schuurman PR, Bosch DA, Bossuyt PM, *et al.* A comparison of continuous thalamic stimulation and thalamotomy for suppression of severe tremor. N Engl J Med 2000; 342(7): 461-8.
[http://dx.doi.org/10.1056/NEJM200002173420703] [PMID: 10675426]

[67]   Nagaseki Y, Shibazaki T, Hirai T, *et al.* Long-term follow-up results of selective VIM-thalamotomy. J Neurosurg 1986; 65(3): 296-302.
[http://dx.doi.org/10.3171/jns.1986.65.3.0296] [PMID: 3734879]

[68]   Selby G. Stereotactic surgery for the relief of Parkinson's disease. 2. An analysis of the results in a series of 303 patients (413 operations). J Neurol Sci 1967; 5(2): 343-75.
[http://dx.doi.org/10.1016/0022-510X(67)90140-2] [PMID: 4862131]

[69]   Matsumoto K, Shichijo F, Fukami T. Long-term follow-up review of cases of Parkinson's disease after unilateral or bilateral thalamotomy. J Neurosurg 1984; 60(5): 1033-44.
[http://dx.doi.org/10.3171/jns.1984.60.5.1033] [PMID: 6716138]

[70]   Boockvar JA, Telfeian A, Baltuch GH, *et al.* Long-term deep brain stimulation in a patient with essential tremor: clinical response and postmortem correlation with stimulator termination sites in ventral thalamus. Case report. J Neurosurg 2000; 93(1): 140-4.
[http://dx.doi.org/10.3171/jns.2000.93.1.0140] [PMID: 10883919]

[71]   Hubble JP, Busenbark KL, Wilkinson S, Penn RD, Lyons K, Koller WC. Deep brain stimulation for essential tremor. Neurology 1996; 46(4): 1150-3.
[http://dx.doi.org/10.1212/WNL.46.4.1150] [PMID: 8780109]

[72]   Koller W, Pahwa R, Busenbark K, *et al.* High-frequency unilateral thalamic stimulation in the treatment of essential and parkinsonian tremor. Ann Neurol 1997; 42(3): 292-9.
[http://dx.doi.org/10.1002/ana.410420304] [PMID: 9307249]

[73]   Pahwa R, Lyons KL, Wilkinson SB, *et al.* Bilateral thalamic stimulation for the treatment of essential tremor. Neurology 1999; 53(7): 1447-50.
[http://dx.doi.org/10.1212/WNL.53.7.1447] [PMID: 10534249]

[74]   Benabid AL, Benazzouz A, Hoffmann D, Limousin P, Krack P, Pollak P. Long-term electrical inhibition of deep brain targets in movement disorders. Mov Disord 1998; 13 (Suppl. 3): 119-25.
[http://dx.doi.org/10.1002/mds.870131321] [PMID: 9827607]

[75]   Niranjan A, Kondziolka D, Baser S, Heyman R, Lunsford LD. Functional outcomes after gamma knife thalamotomy for essential tremor and MS-related tremor. Neurology 2000; 55(3): 443-6.
[http://dx.doi.org/10.1212/WNL.55.3.443] [PMID: 10932286]

[76]   Young RF, Jacques S, Mark R, *et al.* Gamma knife thalamotomy for treatment of tremor: long-term results. J Neurosurg 2000; 93 (Suppl. 3): 128-35.
[PMID: 11143229]

[77]   Siderowf A, Gollump SM, Stern MB, Baltuch GH, Riina HA. Emergence of complex, involuntary movements after gamma knife radiosurgery for essential tremor. Mov Disord 2001; 16(5): 965-7.
[http://dx.doi.org/10.1002/mds.1178] [PMID: 11746633]

[78]   Popa T, Russo M, Vidailhet M, *et al.* Cerebellar rTMS stimulation may induce prolonged clinical benefits in essential tremor, and subjacent changes in functional connectivity: an open label trial. Brain Stimul 2013; 6(2): 175-9.
[http://dx.doi.org/10.1016/j.brs.2012.04.009] [PMID: 22609238]

[79]   Grimaldi G, Argyropoulos GP, Bastian A, *et al.* Cerebellar transcranial direct current stimulation (ctDCS): A novel approach to understanding cerebellar function in health and disease. Neuroscientist

2016; 22(1): 83-97.
[http://dx.doi.org/10.1177/1073858414559409] [PMID: 25406224]

[80]   Helvaci Yilmaz N, Polat B, Hanoglu L. Transcranial direct current stimulation in the treatment of essential tremor: an open-label study. Neurologist 2016; 21(2): 28-9.
[http://dx.doi.org/10.1097/NRL.0000000000000070] [PMID: 26926852]

[81]   Shih LC, Pascual-Leone A. Non-invasive brain stimulation for essential tremor. Tremor Other Hyperkinet Mov (N Y) 2017; 7: 458.
[PMID: 28373927]

[82]   Pfau W, Skog K. Exposure to beta-carbolines norharman and harman. J Chromatogr B Analyt Technol Biomed Life Sci 2004; 802(1): 115-26.
[http://dx.doi.org/10.1016/j.jchromb.2003.10.044] [PMID: 15036003]

[83]   Louis ED, Zheng W. Beta-carboline alkaloids and essential tremor: exploring the environmental determinants of one of the most prevalent neurological diseases. Sci World J 2010; 10: 1783-94.
[http://dx.doi.org/10.1100/tsw.2010.159] [PMID: 20842322]

[84]   Louis ED, Factor-Litvak P, Liu X, *et al.* Elevated brain harmane (1-methyl-9H-pyrido[3,4-b]indole) in essential tremor cases *vs.* controls. Neurotoxicology 2013; 38: 131-5.
[http://dx.doi.org/10.1016/j.neuro.2013.07.002] [PMID: 23911942]

[85]   Kolasiewicz W, Kuter K, Wardas J, Ossowska K. Role of the metabotropic glutamate receptor subtype 1 in the harmaline-induced tremor in rats. J Neural Transm (Vienna) 2009; 116(9): 1059-63.
[http://dx.doi.org/10.1007/s00702-009-0254-5] [PMID: 19551466]

[86]   Louis ED, Keating GA, Bogen KT, Rios E, Pellegrino KM, Factor-Litvak P. Dietary epidemiology of essential tremor: meat consumption and meat cooking practices. Neuroepidemiology 2008; 30(3): 161-6.
[http://dx.doi.org/10.1159/000122333] [PMID: 18382115]

[87]   Li S, Liu W, Teng L, Cheng X, Wang Z, Wang C. Metabolites identification of harmane *in vitro*/*in vivo* in rats by ultra-performance liquid chromatography combined with electrospray ionization quadrupole time-of-flight tandem mass spectrometry. J Pharm Biomed Anal 2014; 92: 53-62.
[http://dx.doi.org/10.1016/j.jpba.2014.01.003] [PMID: 24486683]

[88]   Louis ED, Michalec M, Jiang W, Factor-Litvak P, Zheng W. Elevated blood harmane (1-methyl--H-pyrido[3,4-b]indole) concentrations in Parkinson's disease. Neurotoxicology 2014; 40: 52-6.
[http://dx.doi.org/10.1016/j.neuro.2013.11.005] [PMID: 24300779]

[89]   Louis ED, Benito-León J, Moreno-García S, *et al.* Blood harmane (1-methyl-9H-pyrido[3,4-b]indole) concentration in essential tremor cases in Spain. Neurotoxicology 2013; 34: 264-8.
[http://dx.doi.org/10.1016/j.neuro.2012.09.004] [PMID: 22981972]

[90]   Murkovic M. Chemistry, formation and occurrence of genotoxic heterocyclic aromatic amines in fried products. Eur J Lipid Sci Technol 2004; 106(11): 777-85.
[http://dx.doi.org/10.1002/ejlt.200400993]

[91]   Murkovic M. Formation of heterocyclic aromatic amines in model systems. J Chromatogr B Analyt Technol Biomed Life Sci 2004; 802(1): 3-10.
[http://dx.doi.org/10.1016/j.jchromb.2003.09.026] [PMID: 15035991]

[92]   Rönner B, Lerche H, Bergmüller W, Freilinger C, Severin T, Pischetsrieder M. Formation of tetrahydro-beta-carbolines and beta-carbolines during the reaction of L-tryptophan with D-glucose. J Agric Food Chem 2000; 48(6): 2111-6.
[http://dx.doi.org/10.1021/jf991237l] [PMID: 10888507]

[93]   Herraiz T, Galisteo J. Naturally-occurring tetrahydro-β-carboline alkaloids derived from tryptophan are oxidized to bioactive β-carboline alkaloids by heme peroxidases. Biochem Biophys Res Commun 2014; 451(1): 42-7.
[http://dx.doi.org/10.1016/j.bbrc.2014.07.047] [PMID: 25035927]

[94]   Collins MA, Neafsey EJ. Beta-carboline analogues of N-methyl-4-phenyl-1,2,5,6-tetrahydropyridine
        (MPTP): endogenous factors underlying idiopathic parkinsonism? Neurosci Lett 1985; 55(2): 179-84.
        [http://dx.doi.org/10.1016/0304-3940(85)90016-3] [PMID: 2582318]

[95]   Wernicke C, Schott Y, Enzensperger C, Schulze G, Lehmann J, Rommelspacher H. Cytotoxicity of
        beta-carbolines in dopamine transporter expressing cells: structure-activity relationships. Biochem
        Pharmacol 2007; 74(7): 1065-77.
        [http://dx.doi.org/10.1016/j.bcp.2007.06.046] [PMID: 17692827]

[96]   Storch A, Hwang Y-I, Gearhart DA, *et al.* Dopamine transporter-mediated cytotoxicity of β-
        carbolinium derivatives related to Parkinson's disease: relationship to transporter-dependent uptake. J
        Neurochem 2004; 89(3): 685-94.
        [http://dx.doi.org/10.1111/j.1471-4159.2004.02397.x] [PMID: 15086525]

[97]   Collins MA, Neafsey EJ. β-Carboline Derivatives as Neurotoxins. Birkhäuser BostonPharmacology of
        Endogenous Neurotoxins: A Handbook. Boston, MA 1998; pp. 129-49.

[98]   Matsubara K, Gonda T, Sawada H, *et al.* Endogenously occurring beta-carboline induces parkinsonism
        in nonprimate animals: a possible causative protoxin in idiopathic Parkinson's disease. J Neurochem
        1998; 70(2): 727-35.
        [http://dx.doi.org/10.1046/j.1471-4159.1998.70020727.x] [PMID: 9453568]

[99]   Herraiz T, Guillén H, Galisteo J. N-methyltetrahydro-beta-carboline analogs of 1-methyl-4-phe-
        yl-1,2,3,6-tetrahydropyridine (MPTP) neurotoxin are oxidized to neurotoxic beta-carbolinium cations
        by heme peroxidases. Biochem Biophys Res Commun 2007; 356(1): 118-23.
        [http://dx.doi.org/10.1016/j.bbrc.2007.02.089] [PMID: 17346675]

[100]   Herraiz T, Chaparro C. Human monoamine oxidase enzyme inhibition by coffee and beta-carbolines
        norharman and harman isolated from coffee. Life Sci 2006; 78(8): 795-802.
        [http://dx.doi.org/10.1016/j.lfs.2005.05.074] [PMID: 16139309]

[101]   Wernicke C, Hellmann J, Zięba B, *et al.* 9-Methyl-β-carboline has restorative effects in an animal
        model of Parkinson's disease. Pharmacol Rep 2010; 62(1): 35-53.
        [http://dx.doi.org/10.1016/S1734-1140(10)70241-3] [PMID: 20360614]

[102]   Herraiz T, Chaparro C. Human monoamine oxidase is inhibited by tobacco smoke: beta-carboline
        alkaloids act as potent and reversible inhibitors. Biochem Biophys Res Commun 2005; 326(2): 378-
        86.
        [http://dx.doi.org/10.1016/j.bbrc.2004.11.033] [PMID: 15582589]

[103]   Scott WK, Zhang F, Stajich JM, Scott BL, Stacy MA, Vance JM. Family-based case-control study of
        cigarette smoking and Parkinson disease. Neurology 2005; 64(3): 442-7.
        [http://dx.doi.org/10.1212/01.WNL.0000150905.93241.B2] [PMID: 15699372]

[104]   Emerit J, Edeas M, Bricaire F. Neurodegenerative diseases and oxidative stress. Biomed Pharmacother
        2004; 58(1): 39-46.
        [http://dx.doi.org/10.1016/j.biopha.2003.11.004] [PMID: 14739060]

[105]   Cataldo JK, Prochaska JJ, Glantz SA. Cigarette smoking is a risk factor for Alzheimer's Disease: an
        analysis controlling for tobacco industry affiliation. J Alzheimers Dis 2010; 19(2): 465-80.
        [http://dx.doi.org/10.3233/JAD-2010-1240] [PMID: 20110594]

[106]   Sun P, Zhang S, Li Y, Wang L. Harmine mediated neuroprotection *via* evaluation of glutamate
        transporter 1 in a rat model of global cerebral ischemia. Neurosci Lett 2014; 583: 32-6.
        [http://dx.doi.org/10.1016/j.neulet.2014.09.023] [PMID: 25238961]

[107]   Li Y, Sattler R, Yang EJ, *et al.* Harmine, a natural beta-carboline alkaloid, upregulates astroglial
        glutamate transporter expression. Neuropharmacology 2011; 60(7-8): 1168-75.
        [http://dx.doi.org/10.1016/j.neuropharm.2010.10.016] [PMID: 21034752]

[108]   Allaman I, Bélanger M, Magistretti PJ. Astrocyte-neuron metabolic relationships: for better and for
        worse. Trends Neurosci 2011; 34(2): 76-87.

[http://dx.doi.org/10.1016/j.tins.2010.12.001] [PMID: 21236501]

[109] Błaszczyk JW. Neuropharmacological Review of the Nootropic Herb Bacopa monnieri Rejuvenation Research 2013; 16(4): 313-26.

[110] Buddhala C, Hsu CC, Wu JY. A novel mechanism for GABA synthesis and packaging into synaptic vesicles. Neurochem Int 2009; 55(1-3): 9-12.
[http://dx.doi.org/10.1016/j.neuint.2009.01.020] [PMID: 19428801]

[111] Salazar P, Tapia R. Epilepsy and hippocampal neurodegeneration induced by glutamate decarboxylase inhibitors in awake rats. Epilepsy Res 2015; 116: 27-33.
[http://dx.doi.org/10.1016/j.eplepsyres.2015.06.014] [PMID: 26354164]

[112] Huff CL, Morano RL, Herman JP, Yamamoto BK, Gudelsky GA. MDMA decreases glutamic acid decarboxylase (GAD) 67-immunoreactive neurons in the hippocampus and increases seizure susceptibility: Role for glutamate. Neurotoxicology 2016; 57: 282-90.
[http://dx.doi.org/10.1016/j.neuro.2016.10.011] [PMID: 27773601]

[113] Manto M, Laute MA. A possible mechanism for the beneficial effect of ethanol in essential tremor. Eur J Neurol 2008; 15(7): 697-705.
[http://dx.doi.org/10.1111/j.1468-1331.2008.02150.x] [PMID: 18445025]

[114] Weissmiller AM, Wu C. Current advances in using neurotrophic factors to treat neurodegenerative disorders. Transl Neurodegener 2012; 1(1): 14.
[http://dx.doi.org/10.1186/2047-9158-1-14] [PMID: 23210531]

[115] Kumar GP, Khanum F. Neuroprotective potential of phytochemicals. Pharmacogn Rev 2012; 6(12): 81-90.
[http://dx.doi.org/10.4103/0973-7847.99898] [PMID: 23055633]

[116] Moloudizargari M, Mikaili P, Aghajanshakeri S, Asghari MH, Shayegh J. Pharmacological and therapeutic effects of Peganum harmala and its main alkaloids. Pharmacogn Rev 2013; 7(14): 199-212.
[http://dx.doi.org/10.4103/0973-7847.120524] [PMID: 24347928]

[117] Bensalem S, Soubhye J, Aldib I, *et al.* Inhibition of myeloperoxidase activity by the alkaloids of Peganum harmala L. (Zygophyllaceae). J Ethnopharmacol 2014; 154(2): 361-9.
[http://dx.doi.org/10.1016/j.jep.2014.03.070] [PMID: 24746482]

[118] Herraiz T, Guillén H, Arán VJ, Salgado A. Identification, occurrence and activity of quinazoline alkaloids in Peganum harmala. Food Chem Toxicol 2017; 103: 261-9.
[http://dx.doi.org/10.1016/j.fct.2017.03.010] [PMID: 28279698]

[119] Aguiar S, Borowski T. Neuropharmacological review of the nootropic herb Bacopa monnieri. Rejuvenation Res 2013; 16(4): 313-26.
[http://dx.doi.org/10.1089/rej.2013.1431] [PMID: 23772955]

[120] Ramasamy S, Chin SP, Sukumaran SD, Buckle MJC, Kiew LV, Chung LY. In Silico and *In Vitro* Analysis of Bacoside A Aglycones and Its Derivatives as the Constituents Responsible for the Cognitive Effects of Bacopa monnieri. PLoS One 2015; 10(5): e0126565.
[http://dx.doi.org/10.1371/journal.pone.0126565] [PMID: 25965066]

[121] Krieglstein J, Ausmeier F, El-Abhar H, Lippert K, Welsch M, Rupalla K, *et al.* Neuroprotective effects of Ginkgo biloba constituents. Eur J Pharm Sci 1995; 3(1): 39-48.
[http://dx.doi.org/10.1016/0928-0987(94)00073-9]

[122] Ahlemeyer B, Krieglstein J. Neuroprotective effects of Ginkgo biloba extract. Cell Mol Life Sci 2003; 60(9): 1779-92.
[http://dx.doi.org/10.1007/s00018-003-3080-1] [PMID: 14523543]

[123] Kulkarni R, Girish KJ, Kumar A. Nootropic herbs (Medhya Rasayana) in Ayurveda: An update. Pharmacogn Rev 2012; 6(12): 147-53.
[http://dx.doi.org/10.4103/0973-7847.99949] [PMID: 23055641]

[124] Ramanathan M, Sivakumar S, Anandvijayakumar PR, Saravanababu C, Pandian PR. Neuroprotective evaluation of standardized extract of Centella asciatica in monosodium glutamate treated rats. Indian J Exp Biol 2007; 45(5): 425-31.
[PMID: 17569283]

[125] Verma A, Kumar B, Alam P, Singh V, Gupta SK. RUBIA CORDIFOLIA-A REVIEW ON PHARMACONOSY AND PHYTOCHEMISTRY. Int J Pharm Sci Res 2016; 7(7): 2720.

[126] Abdel Khalik SM, Miyase T, El-Ashaal HA, Melek FR. Triterpenoid saponins from Fagonia cretica. Phytochemistry 2000; 54(8): 853-9.
[http://dx.doi.org/10.1016/S0031-9422(00)00168-0] [PMID: 11014278]

[127] Sivakumar V, Rajan MS. Antioxidant Effect of Tinospora cordifolia Extract in Alloxan-induced Diabetic Rats. Indian J Pharm Sci 2010; 72(6): 795-8.
[http://dx.doi.org/10.4103/0250-474X.84600] [PMID: 21969757]

[128] Rawal AK, Muddeshwar MG, Biswas SK. Rubia cordifolia, Fagonia cretica linn and Tinospora cordifolia exert neuroprotection by modulating the antioxidant system in rat hippocampal slices subjected to oxygen glucose deprivation. BMC Complement Altern Med 2004; 4: 11.
[http://dx.doi.org/10.1186/1472-6882-4-11] [PMID: 15310392]

[129] Venkatesan R, Ji E, Kim SY. Phytochemicals that regulate neurodegenerative disease by targeting neurotrophins: a comprehensive review. BioMed Res Int 2015; 2015: 814068.
[http://dx.doi.org/10.1155/2015/814068] [PMID: 26075266]

[130] Kim DO, Lee KW, Lee HJ, Lee CY. Vitamin C equivalent antioxidant capacity (VCEAC) of phenolic phytochemicals. J Agric Food Chem 2002; 50(13): 3713-7.
[http://dx.doi.org/10.1021/jf020071c] [PMID: 12059148]

[131] Szwajgier D, Borowiec K, Pustelniak K. The Neuroprotective Effects of Phenolic Acids: Molecular Mechanism of Action. Nutrients 2017; 9(5): 477.
[http://dx.doi.org/10.3390/nu9050477] [PMID: 28489058]

[132] Esteban-Fernández A, Rendeiro C, Spencer JPE, del Coso DG, de Llano MDG, Bartolomé B, et al. Neuroprotective Effects of Selected Microbial-Derived Phenolic Metabolites and Aroma Compounds from Wine in Human SH-SY5Y Neuroblastoma Cells and Their Putative Mechanisms of Action Frontiers in Nutrition [Original Research] 2017 March 14; 4(3)

[133] Mandel SA, Avramovich-Tirosh Y, Reznichenko L, et al. Multifunctional activities of green tea catechins in neuroprotection. Modulation of cell survival genes, iron-dependent oxidative stress and PKC signaling pathway. Neurosignals 2005; 14(1-2): 46-60.
[http://dx.doi.org/10.1159/000085385] [PMID: 15956814]

[134] Gutierrez-Merino C, Lopez-Sanchez C, Lagoa R, Samhan-Arias AK, Bueno C, Garcia-Martinez V. Neuroprotective actions of flavonoids. Curr Med Chem 2011; 18(8): 1195-212.
[http://dx.doi.org/10.2174/092986711795029735] [PMID: 21291366]

[135] Janle EM, Lila MA, Grannan M, et al. Pharmacokinetics and tissue distribution of 14C-labeled grape polyphenols in the periphery and the central nervous system following oral administration. J Med Food 2010; 13(4): 926-33.
[http://dx.doi.org/10.1089/jmf.2009.0157] [PMID: 20673061]

[136] Milbury PE, Kalt W. Xenobiotic metabolism and berry flavonoid transport across the blood-brain barrier. J Agric Food Chem 2010; 58(7): 3950-6.
[http://dx.doi.org/10.1021/jf903529m] [PMID: 20128604]

[137] Strathearn KE, Yousef GG, Grace MH, et al. Neuroprotective effects of anthocyanin- and proanthocyanidin-rich extracts in cellular models of Parkinson's disease. Brain Res ;2014 :1555 60-77.
[http://dx.doi.org/10.1016/j.brainres.2014.01.047] [PMID: 24502982]

[138] Richard T, Pawlus AD, Iglésias ML, et al. Neuroprotective properties of resveratrol and derivatives. Ann N Y Acad Sci 2011; 1215: 103-8.

[http://dx.doi.org/10.1111/j.1749-6632.2010.05865.x] [PMID: 21261647]

[139] Ringman JM, Frautschy SA, Cole GM, Masterman DL, Cummings JL. A potential role of the curry spice curcumin in Alzheimer's disease. Curr Alzheimer Res 2005; 2(2): 131-6.
[http://dx.doi.org/10.2174/1567205053585882] [PMID: 15974909]

[140] Cole GM, Teter B, Frautschy SA. Neuroprotective effects of curcumin. Adv Exp Med Biol 2007; 595: 197-212.
[http://dx.doi.org/10.1007/978-0-387-46401-5_8] [PMID: 17569212]

[141] Lapchak PA. Neuroprotective and neurotrophic curcuminoids to treat stroke: a translational perspective. Expert Opin Investig Drugs 2011; 20(1): 13-22.
[http://dx.doi.org/10.1517/13543784.2011.542410] [PMID: 21158690]

CHAPTER 4

The Potential Therapeutic Role of the Melatoninergic System in Treatment of Epilepsy and Comorbid Depression

Jana Tchekalarova[1], Dimitrinka Atanasova[1,2] and Nikolai Lazarov[1,3,*]

[1] *Institute of Neurobiology, Bulgarian Academy of Sciences, Sofia1113, Bulgaria*

[2] *Department of Anatomy, Faculty of Medicine, Trakia University, Stara Zagora6003, Bulgaria*

[3] *Department of Anatomy and Histology, Medical University of Sofia, Sofia1431, Bulgaria*

Abstract: Pharmacoresistant epilepsy is estimated to affect about 30% of patients with epilepsy and predisposes to a higher risk for psychiatric comorbidities and depression. This is one of the most common complications in epilepsy but the mechanisms underlying its development are still unclear. Periclinical studies have shown that selective serotonin reuptake inhibitors (SSRIs) are ineffective against comorbid depression. Dysfunction in circadian rhythms which are driven by the suprachiasmatic nucleus (SCN), is a hallmark of depression. The activity of this circadian pacemaker is under the fine-tuning control of the endogenous hormone melatonin. Over the past decade, there has been extensive research on the therapeutic potential of melatonin and its analogues in the management of both epilepsy and depressive disorders. Melatonin and its analogues targeting the melatonin MT_1 and MT_2 receptors are considered as potential adjuvants for the treatment of epilepsy associated with moderate-to-strong antioxidant, anti-inflammatory, and neuroprotective activity at non-toxic doses. One of the main advantages of the melatonin system is associated with its chronobiotic properties and pivotal role in the resynchronization of disturbed circadian rhythms of different parameters. This chapter summarizes the available experimental and clinical data on melatonin and drugs acting on the MT receptors, which are currently of therapeutic interest in the treatment of epilepsy and depression. Despite the fact that melatonin and drugs based on MT receptors have been used for many purposes over the last three decades, the available data on the potential implementation of melatonin compounds in epilepsy and comorbid depression are scarce. The many unanswered questions regarding the use of melatonin to treat epileptic seizures and complications associated with epilepsy are briefly summarized.

Keywords: Epilepsy, Comorbid depression, Circadian rhythms, Melatonin, Antioxidant, Anti-inflammatory, Neuroprotection, Chronobiotic properties, Treatment.

* **Address for correspondence to Nikolai Lazarov:** Department of Anatomy and Histology, Medical University of Sofia, Sofia 1431, Bulgaria; Tel: +359 2 9172 525; E-mail: nlazarov@medfac.mu-sofia.bg

Atta-ur-Rahman & Zareen Amtul (Eds.)

INRODUCTION

Epilepsy is the fourth most frequent neurological condition and is characterized by recurrent seizures. During seizure onset a pathological and hypersynchronous discharge of clusters of neurons takes place in the brain causing involuntary movements, disturbed emotional and sensational state and, occasionally, loss of awareness. Epileptic patients often suffer of psychiatric complications, especially depression, which represents one of the most common comorbidities leading to worsened quality of life [1, 2]. Several reports considering the link between seizures and depressive behavior suggest a lack of dependence of the comorbid psychiatric state on seizure severity and frequency [3, 4]. Epilepsy has been examined as a risk factor for the appearance of comorbid depression and vice versa thereby confirming the hypothesis that epilepsy and depression have a common underlying pathophysiology [2, 5 - 8]. However, the question about the causal relationship between the two conditions, *i.e.* whether epilepsy leads to depression or vice versa, is still open. Studies exploring comorbid depression reveal that most of the cases in epilepsy patients are atypical or referred to as interictal dysphoric disorder [9 - 12]. For the elaboration of future successful therapies, it is important to understand the underlying mechanisms of comorbid depression in epilepsy which might be different from those of functional depression. Among epilepsy patients, those with temporal lobe epilepsy (TLE) are most frequently diagnosed with comorbid depression. Therefore, TLE models are the most appropriate for exploration of the underlying structural and functional links between the two conditions. In this regard, there is evidence from magnetic resonance imaging that TLE and depression are associated with: 1) common neuronal circuits and connecting pathways, including the temporal, entorhinal, neocortical and frontal lobes with the cingulate gyrus, hippocampus and amygdala, as well as the basal ganglia and thalamus; 2) smaller total brain volumes and/or thinner cortical mantle; 3) hyperactive hypothalamic–pituitary–adrenal (HPA) axis; 4) down-regulated serotonin $(5\text{-HT})_{1A}$ receptors in the mesial frontal, temporal and insular regions, as well as in the raphe nuclei; 5) inflammation (high plasma levels of pro-inflammatory cytokines and gliosis in brain) 6) disrupted balance between excitatory and inhibitory systems [13 - 15].

The majority of currently used anti-epileptic drugs (AEDs) target classical inhibitory GABAergic (*e.g.* phenobarbital, benzodiazepines) and excitatory glutamatergic (Glu) (perampanel) neurotransmitter systems, voltage-gated Na^+-(phenytoin, carbamazepine), or voltage-gated Ca^{2+}-channels (ethosuximide and gabapentin). The necessity to control seizures in severe forms of epilepsy requires high-dosage mono- or polytherapy that is often accompanied by serious side effects, including depression [16]. So far, efficient drugs with disease-modifying capacity which are able to prevent the development of comorbid psychopathology in chronic epilepsy, including depressive state, are lacking on the market. Therefore, understanding the underlying mechanisms either through appropriate animal models or on the basis of clinical data with functional imaging is crucial for the identification and characterization of new effective therapies. In this regard, a substantial amount of information to assist the development of new AEDs, which would be beneficial for complications associated with epilepsy, including comorbid depression, can be obtained from adequate animal models of epilepsy with comorbid depression. The results from such studies would lead to the promotion of promising compounds.

Animal models of TLE associated with depression have already been elaborated by several laboratories [17 - 24]. In different models of epilepsy, animals have demonstrated despair-like behavior in the forced swim test (FST) and anhedonia in a test for preference to sucrose solution (SPT), along with hyperactivity of the HPA axis, inflammation in specific brain structures and in the periphery, structural alterations (atrophy of limbic structures) and functional abnormalities (abnormal theta activity). An intrinsic approach is necessary to explore how early anticonvulsant and antidepressant treatment, even prior to the onset of seizures, will affect the development of epileptogenesis and its concomitant behavioral psychiatric outcome.

The present chapter is focused on preclinical and clinical findings considering the role of the melatonin system in epilepsy and depression, respectively. It also discusses the perspectives for research directed towards the beneficial use of this system against the development of comorbid depression in epilepsy.

PATHOGENESIS OF EPILEPSY

Most epilepsies are caused by brain trauma and are acquired after a latent period of epileptogenesis characterized by neuroplastic reorganization. The general mechanism causing all forms of epilepsy involves an impaired balance between excitatory and inhibitory neurotransmission with dominant excitation. Normally, excitation is under the tonic control of inhibitory interneurons. Different factors such as trauma, genetic mutation, brain tumor, or a number of other complications may result in focal hyperexcitability, which can propagate throughout the central nervous system (CNS) and disrupt the regulatory inhibitory mechanism maintaining a fine-tuning control.

The excitatory/inhibitory equilibrium of the CNS is under the precise control of different modulators and co-transmitters that can affect GABA and Glu neurotransmission through signal transduction [25]. The primary clinical hallmarks in epilepsy are seizures. Mitochondrial dysfunction, oxidative stress, neuroinflammation and plastic neuronal reorganization are believed to play major roles in the pathogenesis of epilepsy [26, 27]. The current targets of AEDs include voltage-gated ion channels, SV2A synaptic vesicle protein, connexins, vesicular Glu and GABA transporters, and Glu and GABA receptors [28]. While cascades of signaling molecules such as metabolic, homeostatic, immune cellular mechanisms and glial cells are not targeted directly, they can contribute to epileptogenesis and result in pharmacoresistance and comorbid complications. Pharmacoresistant epilepsy, which is estimated to affect about 30% of all epilepsy patients, predisposes to a higher risk for psychiatric comorbidities, including depression.

CIRCADIAN RHYTHM ALTERATIONS IN EPILEPSY

Circadian rhythms in mammals are generated by the pacemaker cells of the suprachiasmatic nucleus (SCN) and are also maintained by peripheral circadian oscillators in the hippocampus, liver, lungs, skeletal muscles and other structures. They determine the 24-h cycles of different processes, such as the sleep–wake cycle (SWC), the regulation of core body temperature, blood pressure, and hormonal production. The activity of clock genes, Bmal1 and Clock gene also involving the Period genes (Per1, Per2, Per3), and two Cryptochrome genes (Cry 1, Cry 2), is managed through an intricate autoregulatory translation–transcription feedback mechanism both in the central clock (SCN) and the peripheral clocks [29].

Recently, Matos *et al.* [30] were the first to explore the circadian expressions of clock genes in relation to epilepsy and found that these were significantly disrupted during pilocarpine-induced epileptogenesis. A positive link was established between the phase advance of spontaneous locomotor activity (SLA) and that of the Bmal1 clock gene in the hippocampus in epileptic rats. Matos *et al.* [30] suggest that the impaired phase coupling between the central and peripheral oscillators is triggered by spontaneous recurrent seizures (SRS). Moreover, unlike the Cry1 and Cry2 transcripts, whose rhythmic expression is restored in the chronic epileptic phase, the Per1, Per2, and Per3 transcripts lose permanently their circadian oscillation in epileptic rats suggesting that they play a crucial role in the epileptogenic processes. So far, no data have been reported on clock gene profiles in epilepsy patients.

The SRS are characterized by a 24-h periodicity maintained by an endogenous mechanism in the SCN both in animal models of TLE and in patients with epilepsy [31]. Patients with epilepsy have demonstrated a severe desynchronization of the SWC with abnormal sleep architecture [32], body temperature [31, 33], and SLA [30, 31]. The phase advance in the expression of the clock gene

Bmal1 in the hippocampus is suggested to underlie the phase advance of SLA [30]. Therefore, there might be a common mechanism, which determines the positive correlation between the impaired rhythm of activity of clock genes and different physiological patterns in other peripheral oscillators.

Sleep disturbances can provoke emotional changes with depressive symptoms [34, 35]. Moreover, increased incidence of seizures and treatment with AEDs are able to affect sleep architecture and can worsen the quality of nocturnal sleep [35, 36]. It is accepted that sleep helps the brain to neutralize potentially neurotoxic waste products that are accumulated during wakefulness and to manage the plastic reorganization necessary for normal brain function. Therefore, the improvement of the quality of sleep in patients with epilepsy is a very important objective in the course of medication.

PATHOGENESIS OF DEPRESSION

Depression is currently considered an etiologically heterogeneous disorder due to the inconsistent response rate to commercially available antidepressant drugs and the divergent theories regarding the underlying mechanisms of its onset and development. The monoamine-deficiency theory is the oldest and the most popular theory postulating that depletion of the 5-HT, norepinephrine (NE) and dopamine (DA)- signaling pathways in the CNS plays a central role in the pathophysiology of depression. Revealing that tryptophan is able to enhance the antidepressant effect of a monoamine oxidase inhibitor (MAOI), tranylcypromine, Coppen [37] was the first to propose the monoamine-deficiency theory in 1963. This theory was confirmed later with a successful implementation of the selective 5-HT re-uptake inhibitors (SSRIs) as a first-order choice for the treatment of depression.

The monoamine neurotransmitter systems are distributed diffusely in the brain. They are involved in a variety of behavioral responses including mood, anxiety, sleep, aggression, cognition and locomotion, and another theory suggests that 5-HT and NE mediate different aspects of depression [38]. Furthermore, mono-amine systems can function indirectly through their modulatory action on GABA and Glu neurotransmission. Diminished GABAergic pathway activity is reported in acute and chronic conditions [39, 40]. Meanwhile, the excitatory system is over-activated and drugs targeting Glu neurotransmission such as the N-methy--D-aspartate (NMDA) receptor antagonist ketamine or inhibitors of Glu release alleviate depressive symptoms [39, 41, 42]. Anatomical and functional evidence has revealed diminished activity in specific brain regions, including the left subgenual cingulate cortex, the frontal and temporal cortices, the hippocampus, insula and cerebellum in major depressive disorder [43]. A close link has been postulated to exist between abnormal activity in particular brain structures, decreased neurogenesis and associated neurotrophic factors, excitatory neuro-transmission and HPA axis hyperactivity [39]. Studies on both human and animal models of depression, including comorbid depressive state in models of epilepsy contribute to the progress in the exploration of the basic mechanisms of development of depression that involve the central and peripheral nervous system, and include oxidative stress, synaptic reorganization and abnormal transcription factors [44 - 47].

CIRCADIAN RHYTM ALTERATIONS IN DEPRESSION

Depressive symptoms in unipolar and bipolar forms of depression are associated with disrupted circadian rhythms of behavior (mood, psychomotor activity, memory for positive and negative experiences), physiology (SWC, temperature, metabolism) and endocrine function (cortisol, melatonin) [48]. Both experimental and clinical evidence supports the hypothesis that dysfunction of the molecular mechanisms underlying the circadian clock system, and clock gene polymor-phisms, in particular, is a prerequisite for the development of depression [49 - 52]. Shi *et al.* [53] have recently reported that the link between the genetic polymorphisms of the hClock, hPer3 and hNpas2 clock genes and the predis-position to depression is sex-dependent. In addition, experimental data have demonstrated a significant decrease of BMAL1 and CLOCK protein levels in the prefrontal cortex, which is associated with down-regulation of clock genes such as Pers and Crys in the chronic mild stress model (CMS) of depression in rat [50]. Clinical reports have shown that there is a close link between up-regulated Clock, Period1, and Bmal1 mRNA levels with phase advance or delay in individual circadian genes and the development of depression [51, 52]. All of the abovementioned studies suggest that depression is associated with disruptions in the circadian rhythm of expression of the clock genes.

In general, impaired sleep architecture has been accepted as the most frequent circadian impairment in patients with depression. Treatment of circadian desynchronization, including beneficial approaches such as sleep deprivation, phase advance treatment and light therapy, has revealed promising results for correction of depressive symptoms. Several studies suggest that the abnormal sleep state, characterized with a phase advance of SWC, predisposes to disturbed rhythms in body temperature, hormone release and metabolism [54, 55]. Phase advance in the fluctuations of plasma cortisol/corticosterone, abnormal levels and diurnal patterns of melatonin secretion as well as increased incidence and duration of REM sleep with shortage of slow-wave sleep (SWS) are demonstrated both in patients with depression and in animal models [56 - 61]. Therefore, different hypotheses, associated with circadian changes in depression, have been proposed. One theory postulates that disturbed mood in depression is a consequence of a phase-shift (advance or delay) of SCN and the peripheral oscillators involved in the regulation of temperature, cortisol, melatonin, the SWS and REM sleep, in particular. Thus patients with depression are characterized by disturbed SWC, melatonin secretion and REM sleep expression.

Therapeutic strategies for depression with classical antidepressants target the abnormalities in neurotransmission. Increased incidence and duration of REM sleep is related to diminished SWS and thereby the repair of sleep structure is a crucial component of therapeutic antidepressant strategies. However, in general, MAOIs, SSRIs, and serotonin-norepinephrine reuptake inhibitors (SNRIs) are unable to treat insomnia and about 35% of the patients on this therapy are also prescribed hypnotic drugs to restore sleep architecture [62]. New and rational treatment options for depression must involve correction of disturbed circadian rhythms. In this respect, divergent treatment approaches are implemented such as melatonin administration or bright light exposure. These show beneficial outcomes for resynchronization of the SWC and rhythmic activity of endogenous signaling involved in the regulation of circadian rhythms specifically in patients with "winter depression" [63, 64]. Atypical antidepressants, such as agomelatine, which has both melatonergic and serotonergic receptor action profiles, have shown promising results in experimental animals [58, 60, 66] and in patients with depression [66 - 68], and have an important advantage over classical antidepressants, as they resynchronize the circadian rhythms.

EPILEPSY AND DEPRESSION COMORBIDITY

One of the most frequent comorbidities in epilepsy is depression and there are several excellent reviews summarizing the close relationship between these two conditions [1, 10, 16, 69]. Patients with TLE are reported to be more vulnerable to symptoms of depression and anxiety than those with generalized epilepsy [70]. In pediatric patients, the incidence of depression and anxiety is dependent on the origin of the seizures [71]. Based on preclinical and clinical findings resulting from different approaches, a bidirectional link between the two conditions with a predisposition for appearance of epilepsy in patients with depression and vice versa has been established [6, 7]. Several common neurobiological mechanisms and networks were proposed to underlie this relationship, including the hyperactive HPA axis, neuroinflammation, abnormal activity of 5-HT, norepinephrine, dopamine and Glu and associated anatomical changes in most of the limbic structures and cortical regions [69, 72].

While the focus of therapeutic intervention has been mainly on seizure symptoms in pharmacoresistant epilepsies, the treatment of psychiatric complications often remains underestimated. This issue is additionally complicated by the fact that depressive symptoms encountered in patients with epilepsy are mainly represented by atypical manifestations and are frequently underdiagnosed [72]. Moreover, several AEDs, introduced between 1990 and 2010, which have been effective in the management of intractable forms of epilepsy, have failed to treat or have worsened comorbid psychiatric complications, including depression [16, 73]. Although tricyclic antidepressants as well as SSRI and SNRIs are well tolerated and safe, both experimental and clinical reports have demonstrated that they are not efficient against atypical depressive conditions [21] or may even exacerbate seizure attacks [74]. The lack of activity of fluoxetine against depressive-like responses in post-status epilepticus (SE) rats might be explained with the inability of this SSRI to overcome hippocampal inflammation and attenuate the abnormal activity of the HPA axis [23]. Unterberger *et al.* [75] have recently reported that epileptic patients with well-controlled seizures due to AED therapy do not differ from healthy controls in sleep architecture. Their findings support the hypothesis that both seizure frequency and comorbid psychiatric complications in epilepsy are factors contributing to sleep disturbance.

THE MELATONINERGIC SYSTEM: AN OVERVIEW

Melatonin (*N*-acetyl-5-methoxytryptamine) is mainly secreted from the pineal gland and serves as an important signaling hormone at dark in mammals. It is responsible for the fine-tuning regulation of the circadian rhythms for a variety of biochemical and physiological processes [76 - 78]. Melatonin is also released by the gastrointestinal tract in amounts that allow it to exert a direct protection of its mucosa, the liver and biliary tract [79]. The synchronization performed by the hormone is achieved through adjustment of electrical activity and modulatory control of the main circadian clock, the SCN [80]. The chronotropic function of melatonin is mediated through activation of the G protein-coupled membrane melatonin MT_1 and MT_2 receptors, respectively [81]. The endogenous hormone plays a crucial role in the regulation of sleep rhythms [82] while exogenous melatonin, administered in cases of deficiency of endogenous hormone, induces brain electrical activity similar to sleep state in awake subjects [83]. In addition to its role as a synchronizer, melatonin can exert strong antioxidant and anti-inflammatory effects, as well as neuroprotection [84, 85]. Further to its direct effect as a scavenger of free radicals, the antioxidant activity of this hormone in the brain can be realized through other mechanisms including calmodulin and calreticulin, mitochondrial complex I and the RORα receptors [86, 87]. A deficiency in endogenous melatonin levels is observed with age but also in some diseases with disturbed circadian rhythms, including epilepsy and depression [76,

77, 88]. As a consequence, in humans with impaired rhythms of melatonin release from the pineal gland, the synchronizing control role to many cells in the brain and periphery targeted by the hormone is missing, thereby leading to an impaired circadian rhythm of specific parameters. Because melatonin exerts multiple physiological functions and displays effects throughout the body, its deficiency might trigger numerous devastating processes. Thus, in addition to impaired sleep organization, low plasma levels of melatonin can cause an enhanced vulnerability of cells to oxidative stress, resulting from the normal production of free radicals during metabolism.

Melatonin System and Epilepsy

Preclinical Studies

The anticonvulsant activity of exogenously delivered melatonin is verified in a number of seizure tests in naive rodents summarized in several recent reviews [89 - 91]. Here, we are focusing on the reports considering mainly the role of the melatonin system in models of epilepsy (Table 1). Albertson *et al.* [92] were the first to demonstrate that melatonin is able to attenuate seizure intensity at doses that do not produce sedation in two different kindling models of epilepsy in rats, the pentylenetetrazol (PTZ) and amygdala. These results were later confirmed by Mevissen and Ebert [93] also revealing that daily circadian variations in seizure susceptibility are likely not associated with the diurnal oscillation of melatonin release from the pineal gland. Although melatonin deficit produced by pinealectomy significantly exacerbates the amygdala kindling development [94], the effect of endogenous melatonin on seizure thresholds remains uncertain.

Controversial results have been reported as concerns the effect of the pineal gland hormone on spike and wave discharges (SWDs). While melatonin has been shown to successfully suppress penicillin-induced epileptiform activity [95] and SWDs in the Wistar Albino Glaxo from Rijswijk (WAG/Rij) rat model of absence epilepsy [96] neither pinealectomy nor melatonin supplementation were able to modify the architecture of SWDs in WAG/Rij rats [97]. The discrepancy between the data collected about the genetic form of epilepsy could be explained by the different routes of administration (sub-chronic vs. acute) or the time of recording and detection of absence-like seizures (2 h vs. 24/12 h). Furthermore, opposite results with respect to seizure activity were demonstrated for the melatonin analogue agomelatine, used as an atypical antidepressant in Europe. The results of several acute seizure tests with convulsants with different mechanisms of action in rodents, including WAG/Rij rats, suggest that agomelatin exhibits anticonvulsant activity [96, 98 - 101]. Because of receptor internalization chronic treatment has shown efficiency, in contrast to a single injection [100].

Table 1. Role of melatonin system in animal models of epilepsy.

Animal model	Treatment	Finding	References
Amygdala and pentylenetetrazol kindling rat Kainate (KA)-induced status epilepticus (SE) rat	Melatonin (150-300 mg/kg)	Decrease afterdischarge (AD) length at a non-sedative dose and alleviate seizure severity	Albertson et al. 1981
KA-induced SE rat	Melatonin (2.5 mg/kg x 4 days 0, 1, 2 h after KA)	Decreases seizure severity, neuroprotection in the hippocampus, amygdala and piriform cortex, prevents NA decrease and 5-HIAA increase	Giusti et al. 1996
Amygdala kindling model rat	Melatonin (2.5 mg/kg x 4 days 0, 1, 2 h after KA) Melatonin (75 and 100 mg/kg)	Suppress DNA damage and cell damage in the hippocampus. Raises the current threshold necessary to onset of epileptic ADs; blocks stimulus-induced at threshold generalized seizures	Uz et al. 1996 Mevissen and Ebert 1998
KA-induced SE mice	Melatonin (5 mg/kg before KA)	Alleviates seizure severity, lipid peroxidation and cell damage in CA3 field of the hippocampus	Tan et al. 1998
Pilocarpine-induced SE rat	Melatonin (25 and 50 mg/kg x 7 days before pilocarpine)	Retards SE dose-, time- and age-dependently.	Costa-Lotufo et al. 2002
KA-induced SE mice	Melatonin (20 mg/kg 30 min before, immediately or 15 after KA)	Alleviates seizures during simultaneous injection with KA, suppresses brain mtDNA damages and increased lipid peroxidation Suppresses lipid peroxidation, microglial activation, neuronal damage	Mohanan et al. 2002
KA-induced SE rat	Melatonin (2.5 mg/kg, 20 min before, immediately after KA, 1 and 2 hr after KA)	Melatonin reverses pinealectomy-induced exacerbation of chronic phase	Chung and Han 2003
Pilocarpine-induced model of epilepsy, rats with pinealectomy Amygdala kindling in rat with pinealectomy	Melatonin (2.5 mg/kg 20 min before, concomitantly with pilocarpine, 30 min, 1 h, and 2 h pilocarpine) -	Pinealectomy exacerbated the kindling process Alleviates epileptiform activity without suppression of epilepsy Decreases frequency of spontaneous behavioral seizures, mossy fiber sprouting, and cell loss in chronic phase Suppresses lipid peroxidation in Wistar and SHR rats, reverses activity of antioxidant enzymes to control level, decreases expression of HSP 72 in the hippocampus in Wistar rats Alleviates frequency of spontaneous seizures during treatment, reverse behavioral changes, neuroprotection in the CA1 field of the hippocampus and piriform cortex, raises 5-HT level in the hippocampus Suppresses spontaneous seizures after interruption of treatment, antidepressant effect and neuroprotection in the CA1 field of the hippocampus and piriform cortex Attenuates the genetic absence epilepsy seizures	De Lima et al. 2005

(Table 1) cont......

Animal model	Treatment	Finding	References
Penicillin-induced epileptiform activity rats	Melatonin (20, 40 or 80 µg i.c.v. before penicillin)	Mitigates seizure severity, suppresses neuronal damage in the CA1 field of the hippocampus, restores decline in spatial learning, and LTP impairments, rescues the decreased surface levels of GluR2 in the CA1 region	De Lima et al. 2011
Pilocarpine model of epilepsy rat	Melatonin (30 min, 1, 2, 4, 6, 12, 24, 36, and 48 h after SE onset)	Retards kindling development, alleviates seizures through melatonin receptors; exerts antioxidant effects, corrects despair behavior and liver enzymes.	Atanasova et al. 2013
KA-induced SE Wistar and SHR rats		Enhances spontaneous seizures in the early state of chronic phase, does not corrects behavioral impairments, including impulsivity and in spatial memory but	Tchekalarova et al. 2013
KA model of epilepsy rat	Melatonin (10 mg/kg infusion 14 days before SE)	has strong neuroprotection in the dorsal hippocampus in the CA1, septal CA2, the hilus of the dentate gyrus, piriform cortex and septo-temporal and temporal basolateral amygdala.	Petkova et al. 2014
KA model epilepsy SHR rat	Melatonin (10 mg/kg tap water for two months after SE)	Changes in MT1 and MT2 mRNA and protein expression in 24-hour profile in the hippocampus at different stages of the epileptic process	Aygun et al. 2015
WAG/ Rij rats		Exerts an antidepressant effect, reduces plasma IL-1β levels and suppresses microgliosis and astrogliosis in specific limbic regions	Ma et al. 2017
Pilocarpine model of epilepsy rat	Melatonin (10 mg/kg tap water for two months after SE)	Low levels of melatonin and neuronal loss in the hippocampus in WAG/Rij rats.	Azim et al., 2017
PTZ kindling mice		Melatonin does not affect SWD incidence but has antidepressant effect,	Tchekalarova et al. 2017
KA model of epilepsy rat	Melatonin and Agomelatine (40 mg/kg x 7 days)		Rocha et al. 2017
Pilocarpine model of epilepsy rat	Melatonin (8 mg/kg x 15 day after SE)		Tchekalarova et al. 2018
KA model of epilepsy rat	Agomelatine (10 mg/kg)		Moyanova et al. 2018
WAG/Rij rat model of absence epilepsy and pinealectomy	Agomelatine (40 mg/kg x 30 days after SE)		
	Agomelatine (40 mg/kg x 30 days after SE)	-	
	Melatonin (single injection 80 mg/kg, subchronic 10 mg/kg x 18 days)		

Unlike the effects in acute seizure tests demonstrated by this melatonin analogue in naive rodents, we recently reported that chronic treatment is not only unable to attenuate seizure activity in the kainate (KA) model of TLE, but it decreases the latency for onset of seizures and increases the incidence of SRS between the 2nd and the 3rd week of drug administration in Wistar rats [102]. As concerns melatonin, in general, different laboratories confirm the anticonvulsant activity with respect to SRS of long-term but not of acute treatment, starting before, during or after SE induced by KA or pilocarpine, although the effects depend on age, sex and time of injection [103 - 105]. In addition, several teams have reported that melatonin deficit induced by pinealectomy enhances seizure susceptibility in different models of epilepsy [94, 95, 104]. The plastic changes in the melatonin MT_1 and MT_2 receptors that are observed during the acute, latent and chronic phase of epilepsy are accompanied by altered melatonin levels in the hippocampus of Wistar and WAG/Rij rats [96, 97, 106]. The broadly accepted hypothesis for the anticonvulsant activity of melatonin in models of epilepsy is related to its strong antioxidant and neuroprotective effect against neurotoxicity caused by SE, including elevation of the hydroxyl radicals and severe neuronal loss in vulnerable limbic structures. This assumption is confirmed by all reports considering the effect of melatonin treatment starting during or shortly after SE on lipid peroxidation and mitochondrial and nuclear DNA damage [104, 107 - 114]. While the antioxidant and anti-inflammatory actions of melatonin as well as its analogue agomelatine might contribute to their neuroprotection during SE, it is uncertain whether or not this beneficial activity of the melatonin system is closely associated with the anticonvulsant action observed in clinical trials [115].

Clinical Studies

In general, the studies focused on the effects of melatonin supplementation in patients with epilepsy yield ambiguous results. Several reports reveal a beneficial influence of melatonin therapy in children with myoclonic epilepsy [116] and febrile seizures [117], as well as in patients with intractable epilepsy [118]. However, most of the findings suggest that the beneficial effects of the hormone are associated with its chronotropic activity and its ability to re-synchronize the impaired circadian rhythms in epilepsy, and sleep improvement, in particular, rather than with its anticonvulsant activity. Melatonin treatment is reported to be effective in prevention of seizure attacks mainly in children with sleep problems [119 - 123]. Sleep architecture is normalized resulting in reduced seizure activity in children with intractable epilepsy as well as in patients on valproate therapy [124]. Further, melatonin administration is proposed as an advantageous alternative to the sleep deprivation approach for inducing sleep as it does not to provoke epileptiform discharges and decreases the latency to sleep onset in children [125, 126]. A positive effect of melatonin on oxidative stress is reported in children with epilepsy on valproate and carbamazepine monotherapy [127, 128].

On the other hand, Sandyk *et al.* [88] reveal that melatonin might be a proconvulsant in patients with epilepsy. Seizure exacerbation at night was found to be closely associated with a 5 to 8-fold elevation of the plasma melatonin level in premenstrual and pregnancy periods. Moreover, the authors demonstrated that melatonin is inefficient for improvement of disturbed sleep in epileptic children and it should not be recommended for electroencephalographic (EEG) studies using the sleep deprivation approach.

With few exceptions, the majority of the findings suggests that melatonin can be useful in the treatment of pediatric epilepsy with different etiology due to its positive impact on sleep quality, strong antioxidant effect and lack of toxicity.

THE MELATONINERGIC SYSTEM AND DEPRESSION

Preclinical Studies

There are numerous reports focused on the role of the melatonin system in depression. A summary of the available data on the effects of melatonin and its analogue agomelatine in animal models is presented in Table **2**. The chronic mild model of depression in rats is the most frequently used approach revealing that, compared to fluoxetine, continuous treatment with melatonin has a weaker effect on depressive-like behavior while the melatonin analogue agomelatine shows effects comparable to those of the SSRI [129, 130]. This model is also characterized by impaired circadian rhythms and low plasma melatonin levels [131]. Further, chronic (as opposed to single) administration of melatonin and agomelatine is able to correct depressive-like responses in different models, including a genetic animal model of depression [132], learned helplessness mice [133], a transgenic model for affective disorders with HPA axis feedback control deficit [134], prenatal restraint stress model [135] and rats exposed to chronic constant light [60]. The antidepressant effect of agomelatine is associated with a correction of pathological epigenetic programming [135], induction of neurogenesis [136], improvement in sleep architecture and impaired circadian rhythm of motor activity [58], alleviation of oxidative stress and of inflammation [137]. Agomelatine treatment can exert a chronotropic activity in rats with significant circadian phase advance [65, 66, 68] and this chronotropic effect requires an intact SCN [138]. However, agomelatine is unable to restore the impaired circadian pattern of clock gene expression in CMS suggesting that its antidepressant action is not closely related to molecular circadian rhythms at least in this rat model [139]. Both melatonin and agomelatine treatment can restore the diurnal pattern of the plasma melatonin level in models characterized with depression such as CMS [139, 140] and chronic constant light exposure [60].

Table 2. Role of melatonin system in animal models of depression.

Animal model	Treatment	Finding	References
Flinders Sensitive Line (FSL) rats a genetic animal model of depression Chronic mild stress (CMS) rat	Melatonin agonist S 20304 (1 or 20 mg/kg single and chronic)	Chronic but not single treatment with the highest dose decreases immobility in FST	Overstreet et al. 1998
CMS rat	Melatonin (10 mg/kg x 21 days)	Less effective than fluoxetine to prevent CMS-induced depressive-like response	Kopp et al. 1999
Learned helplessness mice	Melatonin and Agomelatine (10 and 50 mg/kg x 35 days)	Dose-dependent reversal of depressive-like behavior. While agomelatine has comparable to that of imipramine and fluoxetine effect melatonin was less active.	Papp et al. 2003 Bertaina-Anglade et al., 2006
A transgenic model for affective disorders with hypothalamic–pituitary adrenal (HPA) axis feedback control deficit	Melatonin (2, 10 and 50 mg/kg/day x 5 days)	Antidepressant-like activity of agomelatine	Païzanis et al., 2010
Prenatal restraint stress rat CMS rat Prenatal restraint stress rat CMS rat CMS rat CMS rat CMS rat	Agomelatine (10, 50 mg/kg/day x 5 days) Agomelatine (10 mg/kg x 21 days) Agomelatine (40-50 mg/kg x 21-42 days) Agomelatine (40-50 mg/kg x 21-42 days) Agomelatine (Agomelatine (40 mg/kg x 21 days) Agomelatine (2000 ppm x 35 days)	Reverses model-dependent plastic changes and depressive-like behavior through distinct mechanisms compared to classical antidepressants such as fluoxetine Antidepressant effect, corrects pathological epigenetic programming of PRS Corrects CMS-induced decrease in the newborn cell survival in the dentate gyrus	Morley-Fletcher et al. 2011 Dagyt et al., 2011 Mairesse et al. 2013 Stefanovic et al. 2016

(Table 2) cont.....

Animal model	Treatment	Finding	References
Exposure to chronic constant light CMS rat	Agomelatine (40-50 mg/kg x 21-42 days)	Corrects sleep disturbance and the circadian rhythm of motor activity	Demirdaş et al. 2016 Zhao and Fu 2017
	Agomelatine (Agomelatine (40 mg/kg x 21 days) Agomelatine (2000 ppm x 35 days)	Reverses depletion of norepinephrine storage and the mRNA and protein expression of VMAT2, MAO-A protein and mRNA of COMT in the hippocampus.	Sun et al. 2017 Tchekalarova et al. 2018
	Melatonin (10 mg/kg x 28 days)	Suppresses oxidative stress and cytokine production Disrupted circadian rhythm and decreased levels of plasma melatonin levels	Højgaard et al. 2018
	Agomelatine (40 mg/kg x 35 days) -	Rapid onset and prolonged antidepressant effect corrects depressive-like behavior	
	Melatonin (10 mg/kg x 14 days)	through restoration of circadian pattern of plasma melatonin Antidepressant effect and correction in plasma melatonin levels	
	Agomelatine (40 mg/kg x 21 days)	Antidepressant effect through correction of melatonin level	
	Agomelatine (40 mg/kg x 35 days)	during the dark phase but not normalization of disturbed circadian rhythm of clock gene expression	

Clinical Studies

One of the phase-shift hypotheses of depression is related to the critical role of a phase advance or delay of the central pacemaker and associated impairment of the circadian pattern of different parameters, including sleep, temperature and oscillatory secretion of hormones such as cortisol and melatonin. The impairment of SWC and sleep organization in depression is believed to have a serious detrimental influence on mood in patients [54]. Abnormal sleep architecture in depressed patients is associated with shortened latency and increased duration of REM accompanied by a decrease of SWS in comparison with healthy controls [59, 61].

Classical antidepressants, including MAOIs, SSRIs and SNRIs, are reported to cause insomnia [141, 142]. The beneficial effect of melatonin is related mainly with its ability to improve sleep quality but during the day it can worsen depressive symptoms in patients with depression [142]. The resynchronizing role of endogenous melatonin is obscure because the studies focused on this topic provide conflicting results. A disturbed pattern of melatonin release from the pineal gland in patients with depression is reported in several studies [143, 144] but not in others [145].

The development of the new atypical antidepressant agomelatine is related to using a new approach for treatment of the impaired circadian rhythms associated with depression. Thanks to its unique mechanism of action, this drug possesses the advantage of restoring disrupted circadian rhythms through agonistic potency on melatonin receptors in the brain [65, 66, 68]. However, future studies are needed to uncover whether there is a positive correlation between the antidepressant effect and chronotropic activity of agomelatine in patients with depression.

MELATONIN STRATEGY TO TREAT EPILEPSY AND COMORBID DEPRESSION

In search of alternative therapeutic approaches targeting oxidative stress, neuroinflammation, and impaired circadian rhythms with the aim to prevent depressive responses during the chronic phase of epilepsy, we explored the role of the melatonin system in experimental models of comorbid epilepsy and depression. Several models of comorbid epilepsy and depression are validated in WAG/Rij rats with absence epilepsy [147], a pilocarpine model of TLE [21], a kindling model [20], and a rat model of genetic generalized epilepsy [148]. Recently, we reported that the diurnal variability in depressive-like responses in the KA post-SE model of TLE is strain-dependent [24]. While Wistar rats have higher seizure incidence accompanied by anhedonia and increased hopelessness specifically during the light phase, an inactive period for nocturnal animals, spontaneously hypertensive rats (SHRs) show no diurnal fluctuations of depressive-like symptoms which are manifested during both the light and the dark period. Epileptic Wistar rats and SHRs exhibit hippocampal levels of monoamines and tryptophan comparable to those in naive SHRs but lower than the levels in naive Wistar rats. Moreover, naive SHRs are characterized with increased oxidative stress both in the frontal cortex and the hippocampus compared to Wistar rats [107]. A positive correlation has been demonstrated between depressive responses and the frequency of SRS which are more severe in SHRs. A depressive-like behavior is accompanied by a peripheral and central inflammation with a significantly increased plasma level of cytokine IL-1β and gliosis in the hippocampus, basolateral amygdala and piriform cortex in the KA post-SE epileptic Wistar rats (Fig. **1 I, J, K, L**).

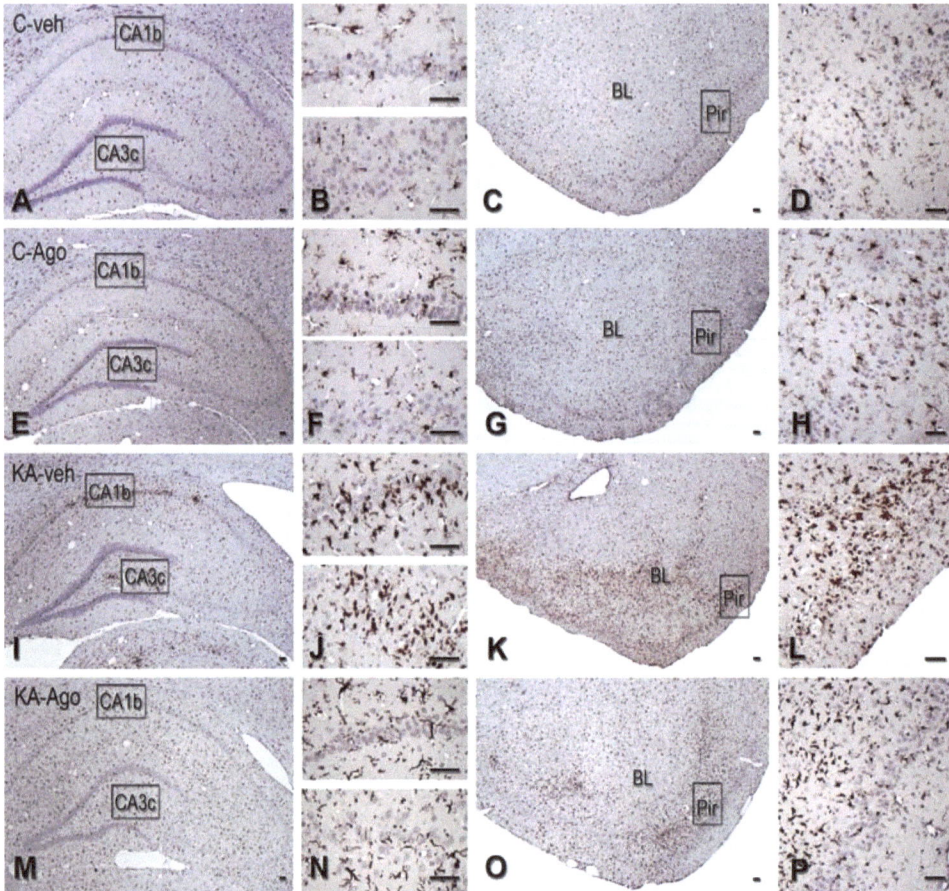

Fig. (1). Immunohistochemical expression of Iba1 protein in the dorsal hippocampus, basolateral amygdala (BL) and piriform cortex (Pir). Immunoreactive microglial cells were observed in two hippocampal cornu ammonis (CA) fields, BL and Pir of **(A, B, C, D)** control rats (C-veh), **(E, F, G, H)** control rats treated with agomelatine (C-Ago), **(I, J, K, L)** epileptic rats (KA-veh) and **(M, N, O, P)** epileptic rats treated with agomelatine (KA-Ago). Activated microglial cells were evident in the epileptic KA-veh group and characterized by round-shaped morphology. Treatment with agomelatine suppressed microgliosis in CA1b, CA3c hippocampal fields, BL and Pir in KA-Ago group **(G, H)**. Scale bars = 50 μm.

Chronic melatonin treatment, starting early after the onset of KA-induced SE, is able to significantly mitigate seizure incidence in Wistar rats only after long-term administration. In addition, melatonin exposure after SE corrects the hippocampal 5-HT level deficit in rats with epilepsy and this effect correlates positively with a suppression of detrimental emotional responses [113]. Unlike in Wistar rats, melatonin administration at the same dose- and time regimen after SE effectively suppresses SRS in SHRs also after discontinuation of treatment. However, melatonin is ineffective against depressive-like responses such as anhedonia and hopelessness in epileptic SHRs [111].

Although melatonin is able to alleviate the increased level of lipid peroxides in the hippocampus both in SHRs and Wistar rats, it enhances the activity of antioxidant enzymes in the hippocampus during KA-induced SE only in Wistar rats pretreated with melatonin [107]. Nevertheless, melatonin fails to prevent seizure severity during SE both in Wistar rats and SHRs suggesting that attenuation of oxidative stress is not enough to mitigate its devastating consequences during epileptogenesis. Unlike melatonin, chronic treatment with agomelatine after KA-induced SE is unable to modify neither the EEG, nor behavioral SRS but can correct hyponeophagia, anhedonia and despair-like responses [115]. These beneficial effects of agomelatine are associated with a strong anti-inflammatory response of the melatonin analogue and suppression of

gliosis, in particular, in the post-SE model of TLE (Fig. **1 M,N,O,P**). Melatonin and agomelatine exert neuroprotection in KA-induced post-SE rats specifically in the CA1 field of the dorsal hippocampus as well as in the piriform cortex (Figs. **2, 3**).

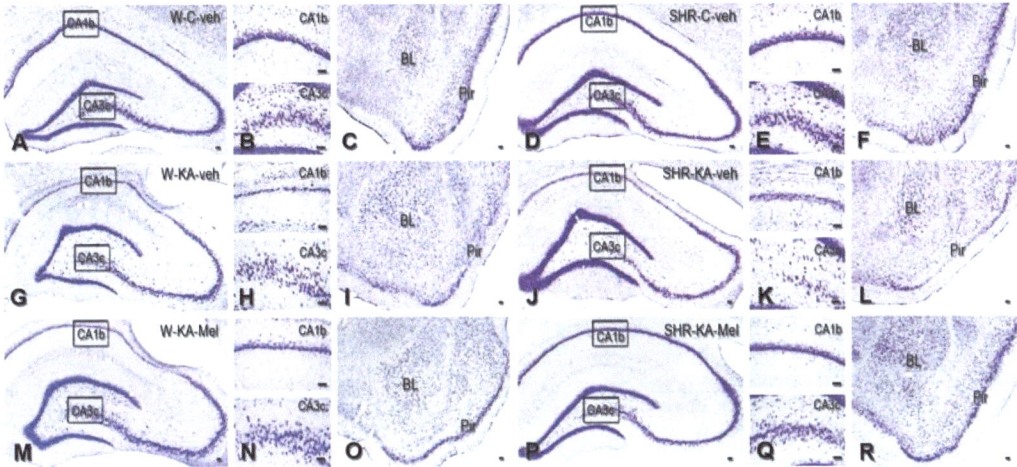

Fig. (2). Nissl-stained coronal sections of the hippocampal formation, BL and Pir of a control rat (**A, B, C, D, E, F**), an epileptic rat treated with a vehicle after SE (**G, I, J, K, L**) and an epileptic rat treated with melatonin after SE (**M, N, O, P, Q, R**). The epileptic rats treated with vehicles demonstrated profound neuronal damage in the CA1 and CA3 field of the dorsal hippocampus (insets in **G**), BL and Pir. Long-term treatment with melatonin during epileptogenesis in the KA post-status epilepticus model of epilepsy exerted neuroprotection in the above-mentioned limbic structures. Scale bars = 200 μm; 50 μm in higher-magnification insets.

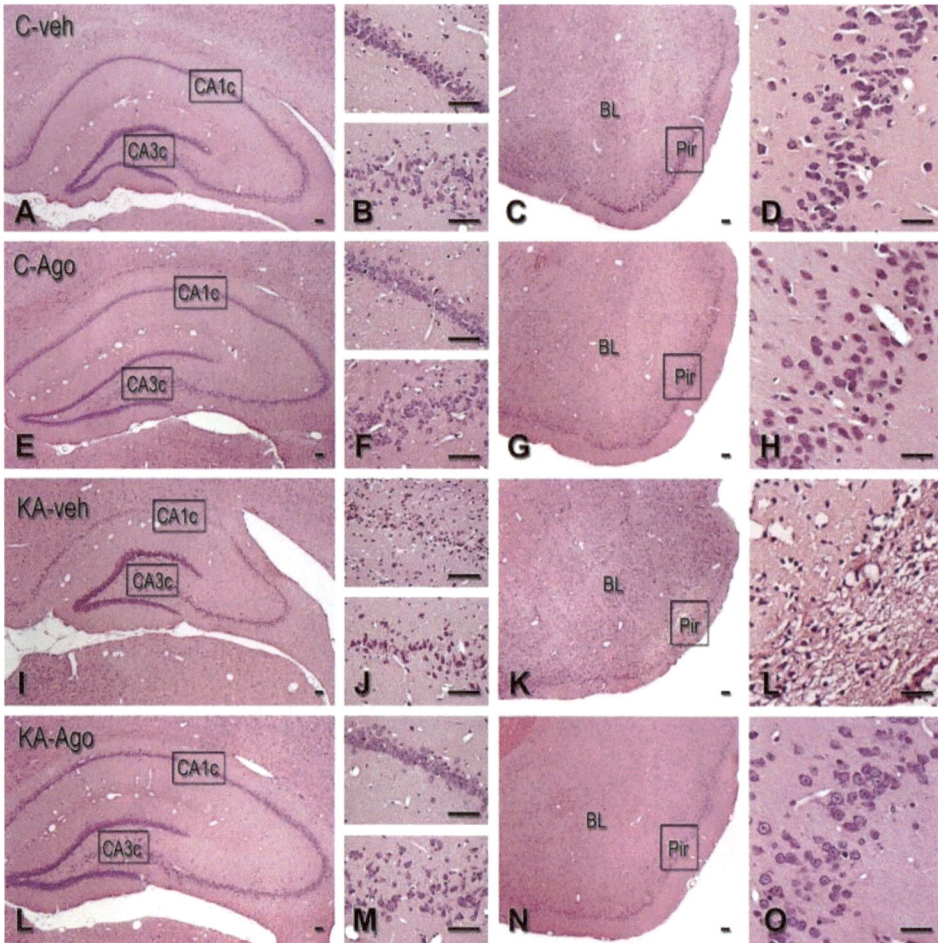

Fig. (3). Hematoxylin & Eosin-stained coronal sections of the hippocampal formation, BL and Pir of a control rat (**A, B, C, D**), an epileptic rat treated with a vehicle after SE (**I, J, K, L**) and an epileptic rat treated with melatonin after SE (**L, M, N, O**). The epileptic rats treated with vehicles demonstrated a severe neuronal damage in the CA1c and CA3c field of the dorsal hippocampus (insets in **I**), BL and Pir. Long-term treatment with agomelatine during epileptogenesis in the KA post-status epilepticus model of epilepsy exerted neuroprotection in the above-mentioned limbic structures. Scale bars = 200 µm; 50 µm in higher-magnification insets.

CONCLUSIONS

The majority of the studies considering the role of the melatonin system in epilepsy and depression support the hypothesis that the activation of this system can mitigate the development of comorbid depressive conditions through the suppression of three devastating processes during epileptogenesis: oxidative stress in SE, inflammation, and neuronal damage in vulnerable limbic regions, the hippocampus, amygdala and piriform cortex. In addition, unlike AEDs and classical antidepressants, compounds targeting the MT receptors have an advantage as chronotropic agents which are able to restore disturbed circadian rhythms both in epilepsy and depression. Even though a number of experimental and clinical studies have shown that the activation of the melatonin system is unable to suppress epileptogenesis and/or the development of chronic epileptic conditions, the development of compounds targeting the melatonin MT receptors is a promising tool for the elaboration of strategies for the treatment of comorbid depression in epilepsy.

AUTHOR CONTRIBUTIONS

J.T.: Conception and design. Collected and interpreted data, wrote the draft; D.A.: Wrote, edited the draft. Design. Prepared panel figures from immunohistochemistry and histology; N.L. gave conceptual advice, interpreted data, contributed to the discussion, reviewed, and wrote and edited the draft. All authors read and approved the final manuscript.

ACKNOWLEDGEMENTS

This work was supported by the National Science Fund of Bulgaria (research grants No. DN 03/10, 2016; No. DN-12/6, 2017; No. DM 11/4, 2017) and the Medical University of Sofia, Bulgaria (grant No. D66/2018). We would like to thank Elena Petrova for professional editing of the manuscript.

CONFLICT OF INTEREST

The authors declare no conflict of interest

ABBREVIATIONS

TLE	Temporal lobe epilepsy
HPA	hypothalamic–pituitary–adrenal
5-HT	serotonin
AEDs	anti-epileptic drugs
Glu	glutamatergic
FST	forced swim test
SPT	test for preference to sucrose solution
SCN	suprachiasmatic nucleus

SWC	sleep–wake cycle
SRS	spontaneous recurrent seizures
SLA	spontaneous locomotor activity
NE	norepinephrine
DA	dopamine
MAOI	monoamine oxidase inhibitor
SSRIs	selective serotonin re-uptake inhibitors
NMDA	N-methyl-D-aspartate
SWS	slow-wave sleep
SNRIs	serotonin-norepinephrine reuptake inhibitors
CMS	chronic mild stress model
SE	status epilepticus
PTZ	pentylenetetrazol
SWDs	spike and spike-wave discharges
KA	kainate
SHRs	spontaneously hypertensive rats
EEG	electroencephalography

REFERENCES

[1]   Błaszczyk B, Czuczwar SJ. Epilepsy coexisting with depression. Pharmacol Rep 2016; 68(5): 1084-92.
[http://dx.doi.org/10.1016/j.pharep.2016.06.011] [PMID: 27634589]

[2]   Kondziella D, Alvestad S, Vaaler A, Sonnewald U. Which clinical and experimental data link temporal lobe epilepsy with depression? J Neurochem 2007; 103(6): 2136-52.
[http://dx.doi.org/10.1111/j.1471-4159.2007.04926.x] [PMID: 17887964]

[3]   Gilliam FG, Maton BM, Martin RC, *et al.* Hippocampal 1H-MRSI correlates with severity of depression symptoms in temporal lobe epilepsy. Neurology 2007; 68(5): 364-8.
[http://dx.doi.org/10.1212/01.wnl.0000252813.86812.81] [PMID: 17261683]

[4]   Spencer SS, Berg AT, Vickrey BG, *et al.* Initial outcomes in the multicenter study of epilepsy surgery. Neurology 2003; 61(12): 1680-5.
[http://dx.doi.org/10.1212/01.WNL.0000098937.35486.A3] [PMID: 14694029]

[5]   Forsgren L, Nyström L. An incident case-referent study of epileptic seizures in adults. Epilepsy Res 1990; 6(1): 66-81.
[http://dx.doi.org/10.1016/0920-1211(90)90010-S] [PMID: 2357957]

[6]   Hesdorffer DC, Hauser WA, Annegers JF, Cascino G. Major depression is a risk factor for seizures in older adults. Ann Neurol 2000; 47(2): 246-9.
[http://dx.doi.org/10.1002/1531-8249(200002)47:2<246::AID-ANA17>3.0.CO;2-E]   [PMID: 10665498]

[7]   Hesdorffer DC, Lúdvígsson P, Hauser WA, Olafsson E, Kjartansson O. Co-occurrence of major depression or suicide attempt with migraine with aura and risk for unprovoked seizure. Epilepsy Res 2007; 75(2-3): 220-3.
[http://dx.doi.org/10.1016/j.eplepsyres.2007.05.001] [PMID: 17572070]

[8]   Tellez-Zenteno JF, Patten SB, Jetté N, Williams J, Wiebe S. Psychiatric comorbidity in epilepsy: a population-based analysis. Epilepsia 2007; 48(12): 2336-44.
[http://dx.doi.org/10.1111/j.1528-1167.2007.01222.x] [PMID: 17662062]

[9]   Blumer D. Dysphoric disorders and paroxysmal affects: recognition and treatment of epilepsy-related psychiatric disorders. Harv Rev Psychiatry 2000; 8(1): 8-17.
[http://dx.doi.org/10.3109/hrp.8.1.8] [PMID: 10824293]

[10]   Kanner AM. Depression and epilepsy: a new perspective on two closely related disorders. Epilepsy Curr 2006; 6(5): 141-6.
[http://dx.doi.org/10.1111/j.1535-7511.2006.00125.x] [PMID: 17260039]

[11]   Mendez MF, Cummings JL, Benson DF. Depression in epilepsy. Significance and phenomenology. Arch Neurol 1986; 43(8): 766-70.
[http://dx.doi.org/10.1001/archneur.1986.00520080014012] [PMID: 3729756]

[12]   Zobel A, Wellmer J, Schulze-Rauschenbach S, *et al.* Impairment of inhibitory control of the hypothalamic pituitary adrenocortical system in epilepsy. Eur Arch Psychiatry Clin Neurosci 2004; 254(5): 303-11.
[http://dx.doi.org/10.1007/s00406-004-0499-9] [PMID: 15365705]

[13]   Cascino GD, Jack CR Jr, Parisi JE, *et al.* Magnetic resonance imaging-based volume studies in temporal lobe epilepsy: pathological correlations. Ann Neurol 1991; 30(1): 31-6.
[http://dx.doi.org/10.1002/ana.410300107] [PMID: 1929226]

[14]   Kanner AM. Is depression associated with an increased risk of treatment-resistant epilepsy? Research strategies to investigate this question. Epilepsy Behav 2014; 38: 3-7.
[http://dx.doi.org/10.1016/j.yebeh.2014.06.027] [PMID: 25260238]

[15]   Valente KD, Busatto Filho G. Depression and temporal lobe epilepsy represent an epiphenomenon sharing similar neural networks: clinical and brain structural evidences. Arq Neuropsiquiatr 2013; 71(3): 183-90.
[http://dx.doi.org/10.1590/S0004-282X2013000300011] [PMID: 23563720]

[16]   Kanner AM, Schachter SC, Barry JJ, *et al.* Depression and epilepsy: epidemiologic and neurobiologic perspectives that may explain their high comorbid occurrence. Epilepsy Behav 2012; 24(2): 156-68.
[http://dx.doi.org/10.1016/j.yebeh.2012.01.007] [PMID: 22632406]

[17]   Epps SA, Weinshenker D. Rhythm and blues: animal models of epilepsy and depression comorbidity. Biochem Pharmacol 2013; 85(2): 135-46.
[http://dx.doi.org/10.1016/j.bcp.2012.08.016] [PMID: 22940575]

[18]   Citraro R, Scicchitano F, De Fazio S, *et al.* Preclinical activity profile of α-lactoalbumin, a whey protein rich in tryptophan, in rodent models of seizures and epilepsy. Epilepsy Res 2011; 95(1-2): 60-9.
[http://dx.doi.org/10.1016/j.eplepsyres.2011.02.013] [PMID: 21458955]

[19]   Kalynchuk LE, Pinel JP, Treit D. Long-term kindling and interictal emotionality in rats: effect of stimulation site. Brain Res 1998; 779(1-2): 149-57.
[http://dx.doi.org/10.1016/S0006-8993(97)01110-4] [PMID: 9473643]

[20]   Mazarati A, Shin D, Auvin S, Caplan R, Sankar R. Kindling epileptogenesis in immature rats leads to persistent depressive behavior. Epilepsy Behav 2007; 10(3): 377-83.
[http://dx.doi.org/10.1016/j.yebeh.2007.02.001] [PMID: 17368107]

[21]   Mazarati A, Siddarth P, Baldwin RA, Shin D, Caplan R, Sankar R. Depression after status epilepticus: behavioural and biochemical deficits and effects of fluoxetine. Brain 2008; 131(Pt 8): 2071-83.
[http://dx.doi.org/10.1093/brain/awn117] [PMID: 18559371]

[22]   Mortazavi F, Ericson M, Story D, Hulce VD, Dunbar GL. Spatial learning deficits and emotional impairments in pentylenetetrazole-kindled rats. Epilepsy Behav 2005; 7(4): 629-38.
[http://dx.doi.org/10.1016/j.yebeh.2005.08.019] [PMID: 16246633]

[23]   Pineda E, Shin D, Sankar R, Mazarati AM. Comorbidity between epilepsy and depression: experimental evidence for the involvement of serotonergic, glucocorticoid, and neuroinflammatory mechanisms. Epilepsia 2010; 51 (Suppl. 3): 110-4.
[http://dx.doi.org/10.1111/j.1528-1167.2010.02623.x] [PMID: 20618414]

[24]   Tchekalarova J, Pechlivanova D, Atanasova T, Markova P, Lozanov V, Stoynev A. Diurnal variations in depression-like behavior of Wistar and spontaneously hypertensive rats in the kainate model of temporal lobe epilepsy. Epilepsy Behav 2011; 20(2): 277-85.
[http://dx.doi.org/10.1016/j.yebeh.2010.12.021] [PMID: 21277833]

[25]   L Devlin A, Odell M, L Charlton J, Koppel S. Epilepsy and driving: current status of research. Epilepsy Res 2012; 102(3): 135-52.
[http://dx.doi.org/10.1016/j.eplepsyres.2012.08.003] [PMID: 22981339]

[26]   Kovac S, Dinkova Kostova AT, Herrmann AM, Melzer N, Meuth SG, Gorji A. Metabolic and homeostatic changes in seizures and acquired epilepsy – mitochondria, calcium dynamics and reactive oxygen species. Int J Mol Sci 2017; 18(9): 1935.
[http://dx.doi.org/10.3390/ijms18091935] [PMID: 28885567]

[27]   Vezzani A. Epilepsy and inflammation in the brain: overview and pathophysiology. Epilepsy Curr 2014; 14(1) (Suppl.): 3-7.
[http://dx.doi.org/10.5698/1535-7511-14.s2.3] [PMID: 24955068]

[28]   Voskuyl R, Clinckers R. Pharmacological approaches for the assessment of anti-epileptic drug efficacy in experimental animal models.Encyclopedia of Basic Research in Epilepsy. Elsevier 2009.
[http://dx.doi.org/10.1016/B978-012373961-2.00235-6]

[29]   Mohawk JA, Green CB, Takahashi JS. Central and peripheral circadian clocks in mammals. Annu Rev Neurosci 2012; 35: 445-62.
[http://dx.doi.org/10.1146/annurev-neuro-060909-153128] [PMID: 22483041]

[30]   Matos HC, Koike BDV, Pereira WDS, et al. Rhythms of core clock genes and spontaneous locomotor activity in post-status epilepticus model of mesial temporal lobe epilepsy. Front Neurol 2018; 9: 632.
[http://dx.doi.org/10.3389/fneur.2018.00632] [PMID: 30116220]

[31]   Quigg M, Clayburn H, Straume M, Menaker M, Bertram EH III. Effects of circadian regulation and rest-activity state on spontaneous seizures in a rat model of limbic epilepsy. Epilepsia 2000; 41(5): 502-9.
[http://dx.doi.org/10.1111/j.1528-1157.2000.tb00202.x] [PMID: 10802754]

[32]   Foldvary-Schaefer N, Grigg-Damberger M. Sleep and epilepsy: what we know, don't know, and need to know. J Clin Neurophysiol 2006; 23(1): 4-20.
[http://dx.doi.org/10.1097/01.wnp.0000206877.90232.cb] [PMID: 16514348]

[33]   Quigg M, Clayburn H, Straume M, Menaker M, Bertram EH III. Hypothalamic neuronal loss and altered circadian rhythm of temperature in a rat model of mesial temporal lobe epilepsy. Epilepsia 1999; 40(12): 1688-96.
[http://dx.doi.org/10.1111/j.1528-1157.1999.tb01585.x] [PMID: 10612331]

[34]   Fountoulakis KN. Disruption of biological rhythms as a core problem and therapeutic target in mood disorders: the emerging concept of 'rhythm regulators'. Ann Gen Psychiatry 2010; 9: 3.
[http://dx.doi.org/10.1186/1744-859X-9-3] [PMID: 20157624]

[35]   Wulff K, Gatti S, Wettstein JG, Foster RG. Sleep and circadian rhythm disruption in psychiatric and neurodegenerative disease. Nat Rev Neurosci 2010; 11(8): 589-99.
[http://dx.doi.org/10.1038/nrn2868] [PMID: 20631712]

[36]   Hofstra WA, de Weerd AW. The circadian rhythm and its interaction with human epilepsy: a review of literature. Sleep Med Rev 2009; 13(6): 413-20.
[http://dx.doi.org/10.1016/j.smrv.2009.01.002] [PMID: 19398353]

[37]   Coppen A, Shaw DM, Farrell JP. Potentiation of the antidepressive effect of a monoamine-oxidase

inhibitor by tryptophan. Lancet 1963; 1(7272): 79-81.
[http://dx.doi.org/10.1016/S0140-6736(63)91084-5] [PMID: 14022907]

[38] Healy D. The Antidepressant Era. Cambridge: Harvard University Press 1997; pp. 1-317.

[39] Hasler G. Pathophysiology of depression: do we have any solid evidence of interest to clinicians? World Psychiatry 2010; 9(3): 155-61.
[http://dx.doi.org/10.1002/j.2051-5545.2010.tb00298.x] [PMID: 20975857]

[40] Rajkowska G, O'Dwyer G, Teleki Z, Stockmeier CA, Miguel-Hidalgo JJ. GABAergic neurons immunoreactive for calcium binding proteins are reduced in the prefrontal cortex in major depression. Neuropsychopharmacology 2007; 32(2): 471-82.
[http://dx.doi.org/10.1038/sj.npp.1301234] [PMID: 17063153]

[41] Kendell SF, Krystal JH, Sanacora G. GABA and glutamate systems as therapeutic targets in depression and mood disorders. Expert Opin Ther Targets 2005; 9(1): 153-68.
[http://dx.doi.org/10.1517/14728222.9.1.153] [PMID: 15757488]

[42] Zarate CA Jr, Singh JB, Carlson PJ, *et al.* A randomized trial of an N-methyl-D-aspartate antagonist in treatment-resistant major depression. Arch Gen Psychiatry 2006; 63(8): 856-64.
[http://dx.doi.org/10.1001/archpsyc.63.8.856] [PMID: 16894061]

[43] Fitzgerald PB, Laird AR, Maller J, Daskalakis ZJ. A meta-analytic study of changes in brain activation in depression. Hum Brain Mapp 2008; 29(6): 683-95.
[http://dx.doi.org/10.1002/hbm.20426] [PMID: 17598168]

[44] Capuron L, Miller AH. Immune system to brain signaling: neuropsychopharmacological implications. Pharmacol Ther 2011; 130(2): 226-38.
[http://dx.doi.org/10.1016/j.pharmthera.2011.01.014] [PMID: 21334376]

[45] Delpech JC, Madore C, Nadjar A, Joffre C, Wohleb ES, Layé S. Microglia in neuronal plasticity: Influence of stress. Neuropharmacology 2015; 96(Pt A): 19-28.
[http://dx.doi.org/10.1016/j.neuropharm.2014.12.034] [PMID: 25582288]

[46] Maes M, Kubera M, Obuchowiczwa E, Goehler L, Brzeszcz J. Depression's multiple comorbidities explained by (neuro)inflammatory and oxidative & nitrosative stress pathways. Neuroendocrinol Lett 2011; 32(1): 7-24.
[PMID: 21407167]

[47] Malki K, Keers R, Tosto MG, *et al.* The endogenous and reactive depression subtypes revisited: integrative animal and human studies implicate multiple distinct molecular mechanisms underlying major depressive disorder. BMC Med 2014; 12: 73.
[http://dx.doi.org/10.1186/1741-7015-12-73] [PMID: 24886127]

[48] Wirz-Justice A. Biological rhythms in mood disorders.Psychopharmacology: the Fourth Generation of Progress. New York: Raven Press 1995; pp. 999-1017.

[49] Benedetti F, Serretti A, Colombo C, *et al.* Influence of CLOCK gene polymorphism on circadian mood fluctuation and illness recurrence in bipolar depression. Am J Med Genet B Neuropsychiatr Genet 2003; 123B(1): 23-6.
[http://dx.doi.org/10.1002/ajmg.b.20038] [PMID: 14582141]

[50] Calabrese F, Savino E, Papp M, Molteni R, Riva MA. Chronic mild stress-induced alterations of clock gene expression in rat prefrontal cortex: modulatory effects of prolonged lurasidone treatment. Pharmacol Res 2016; 104: 140-50.
[http://dx.doi.org/10.1016/j.phrs.2015.12.023] [PMID: 26742719]

[51] Gouin J-P, Connors J, Kiecolt-Glaser JK, *et al.* Altered expression of circadian rhythm genes among individuals with a history of depression. J Affect Disord 2010; 126(1-2): 161-6.
[http://dx.doi.org/10.1016/j.jad.2010.04.002] [PMID: 20471092]

[52] Li JZ, Bunney BG, Meng F, *et al.* Circadian patterns of gene expression in the human brain and disruption in major depressive disorder. Proc Natl Acad Sci USA 2013; 110(24): 9950-5.

[http://dx.doi.org/10.1073/pnas.1305814110] [PMID: 23671070]

[53]   Shi SQ, White MJ, Borsetti HM, *et al.* Molecular analyses of circadian gene variants reveal sex-dependent links between depression and clocks. Transl Psychiatry 2016; 6e748
[http://dx.doi.org/10.1038/tp.2016.9] [PMID: 26926884]

[54]   Germain A, Kupfer DJ. Circadian rhythm disturbances in depression. Hum Psychopharmacol 2008; 23(7): 571-85.
[http://dx.doi.org/10.1002/hup.964] [PMID: 18680211]

[55]   Soria V, Urretavizcaya M. [Circadian rhythms and depression]. Actas Esp Psiquiatr 2009; 37(4): 222-32.
[PMID: 19927234]

[56]   Káradóttir R, Axelsson J. Melatonin secretion in SAD patients and healthy subjects matched with respect to age and sex. Int J Circumpolar Health 2001; 60(4): 548-51.
[PMID: 11768433]

[57]   Koenigsberg HW, Teicher MH, Mitropoulou V, *et al.* 24-h Monitoring of plasma norepinephrine, MHPG, cortisol, growth hormone and prolactin in depression. J Psychiatr Res 2004; 38(5): 503-11.
[http://dx.doi.org/10.1016/j.jpsychires.2004.03.006] [PMID: 15380401]

[58]   Mairesse J, Silletti V, Laloux C, *et al.* Chronic agomelatine treatment corrects the abnormalities in the circadian rhythm of motor activity and sleep/wake cycle induced by prenatal restraint stress in adult rats. Int J Neuropsychopharmacol 2013; 16(2): 323-38.
[http://dx.doi.org/10.1017/S1461145711001970] [PMID: 22310059]

[59]   Shaffery J, Hoffmann R, Armitage R. The neurobiology of depression: perspectives from animal and human sleep studies. Neuroscientist 2003; 9(1): 82-98.
[http://dx.doi.org/10.1177/1073858402239594] [PMID: 12580343]

[60]   Tchekalarova J, Stoynova T, Ilieva K, Mitreva R, Atanasova M. Agomelatine treatment corrects symptoms of depression and anxiety by restoring the disrupted melatonin circadian rhythms of rats exposed to chronic constant light. Pharmacol Biochem Behav 2018; 171: 1-9.
[http://dx.doi.org/10.1016/j.pbb.2018.05.016] [PMID: 29807067]

[61]   Tsuno N, Besset A, Ritchie K. Sleep and depression. J Clin Psychiatry 2005; 66(10): 1254-69.
[http://dx.doi.org/10.4088/JCP.v66n1008] [PMID: 16259539]

[62]   Rascati K. Drug utilization review of concomitant use of specific serotonin reuptake inhibitors or clomipramine with antianxiety/sleep medications. Clin Ther 1995; 17(4): 786-90.
[http://dx.doi.org/10.1016/0149-2918(95)80055-7] [PMID: 8565041]

[63]   Lam RW, Levitan RD. Pathophysiology of seasonal affective disorder: a review. J Psychiatry Neurosci 2000; 25(5): 469-80.
[PMID: 11109298]

[64]   Lewy AJ, Bauer VK, Cutler NL, *et al.* Morning vs evening light treatment of patients with winter depression. Arch Gen Psychiatry 1998; 55(10): 890-6.
[http://dx.doi.org/10.1001/archpsyc.55.10.890] [PMID: 9783559]

[65]   Weibel L, Turek FW, Mocaer E, Van Reeth O. A melatonin agonist facilitates circadian resynchronization in old hamsters after abrupt shifts in the light-dark cycle. Brain Res 2000; 880(1-2): 207-11.
[http://dx.doi.org/10.1016/S0006-8993(00)02806-7] [PMID: 11033009]

[66]   Kräuchi K, Cajochen C, Möri D, Graw P, Wirz-Justice A. Early evening melatonin and S-20098 advance circadian phase and nocturnal regulation of core body temperature. Am J Physiol 1997; 272(4 Pt 2): R1178-88.
[PMID: 9140018]

[67]   San L, Arranz B. Agomelatine: a novel mechanism of antidepressant action involving the melatonergic and the serotonergic system. Eur Psychiatry 2008; 23(6): 396-402.

[http://dx.doi.org/10.1016/j.eurpsy.2008.04.002] [PMID: 18583104]

[68]   Van Reeth O, Weibel L, Olivares E, Maccari S, Mocaer E, Turek FW. Melatonin or a melatonin agonist corrects age-related changes in circadian response to environmental stimulus. Am J Physiol Regul Integr Comp Physiol 2001; 280(5): R1582-91.
       [http://dx.doi.org/10.1152/ajpregu.2001.280.5.R1582] [PMID: 11294784]

[69]   Sankar R, Mazarati A. Neurobiology of depression as a comorbidity of epilepsy.Jasper's Basic Mechanisms of the Epilepsies. 4th ed. Bethesda, MD: National Center for Biotechnology Information, Oxford University Press, USA 2012; pp. 1399-416.
       [http://dx.doi.org/10.1093/med/9780199746545.003.0074]

[70]   Piazzini A, Canevini MP, Maggiori G, Canger R. Depression and anxiety in patients with epilepsy. Epilepsy Behav 2001; 2(5): 481-9.
       [http://dx.doi.org/10.1006/ebeh.2001.0247] [PMID: 12609287]

[71]   Caplan R, Siddarth P, Gurbani S, Hanson R, Sankar R, Shields WD. Depression and anxiety disorders in pediatric epilepsy. Epilepsia 2005; 46(5): 720-30.
       [http://dx.doi.org/10.1111/j.1528-1167.2005.43604.x] [PMID: 15857439]

[72]   Kanner AM. Depression in epilepsy: a neurobiologic perspective. Epilepsy Curr 2005; 5(1): 21-7.
       [http://dx.doi.org/10.1111/j.1535-7597.2005.05106.x] [PMID: 16059450]

[73]   Naess S, Eriksen J, Tambs K. Psychological well-being of people with epilepsy in Norway. Epilepsy Behav 2007; 11(3): 310-5.
       [http://dx.doi.org/10.1016/j.yebeh.2007.06.004] [PMID: 17825627]

[74]   Kanner AM. The treatment of depressive disorders in epilepsy: what all neurologists should know. Epilepsia 2013; 54 (Suppl. 1): 3-12.
       [http://dx.doi.org/10.1111/epi.12100] [PMID: 23458461]

[75]   Unterberger I, Gabelia D, Prieschl M, *et al.* Sleep disorders and circadian rhythm in epilepsy revisited: a prospective controlled study. Sleep Med 2015; 16(2): 237-42.
       [http://dx.doi.org/10.1016/j.sleep.2014.09.021] [PMID: 25637104]

[76]   Schapel GJ, Beran RG, Kennaway DL, McLoughney J, Matthews CD. Melatonin response in active epilepsy. Epilepsia 1995; 36(1): 75-8.
       [http://dx.doi.org/10.1111/j.1528-1157.1995.tb01669.x] [PMID: 8001514]

[77]   Hardeland R. Melatonin and synthetic melatoninergic agonists in psychiatric and age-associated disorders: successful and unsuccessful approaches. Curr Pharm Des 2016; 22(8): 1086-101.
       [http://dx.doi.org/10.2174/1381612822666151214125543] [PMID: 25248806]

[78]   Herxheimer A, Petrie KJ. Melatonin for the prevention and treatment of jet lag. Cochrane Database Syst Rev 2002; 2(2)CD001520
       [http://dx.doi.org/10.1002/14651858.CD001520] [PMID: 12076414]

[79]   Reiter RJ, Rosales-Corral S, Coto-Montes A, *et al.* The photoperiod, circadian regulation and chronodisruption: the requisite interplay between the suprachiasmatic nuclei and the pineal and gut melatonin. J Physiol Pharmacol 2011; 62(3): 269-74.
       [PMID: 21893686]

[80]   Arendt J, Skene DJ. Melatonin as a chronobiotic. Sleep Med Rev 2005; 9(1): 25-39.
       [http://dx.doi.org/10.1016/j.smrv.2004.05.002] [PMID: 15649736]

[81]   Pandi-Perumal SR, Trakht I, Srinivasan V, *et al.* Physiological effects of melatonin: role of melatonin receptors and signal transduction pathways. Prog Neurobiol 2008; 85(3): 335-53.
       [http://dx.doi.org/10.1016/j.pneurobio.2008.04.001] [PMID: 18571301]

[82]   Cajochen C, Kräuchi K, Wirz-Justice A. Role of melatonin in the regulation of human circadian rhythms and sleep. J Neuroendocrinol 2003; 15(4): 432-7.
       [http://dx.doi.org/10.1046/j.1365-2826.2003.00989.x] [PMID: 12622846]

[83]   Gorfine T, Assaf Y, Goshen-Gottstein Y, Yeshurun Y, Zisapel N. Sleep-anticipating effects of melatonin in the human brain. Neuroimage 2006; 31(1): 410-8.
[http://dx.doi.org/10.1016/j.neuroimage.2005.11.024] [PMID: 16427787]

[84]   Pandi-Perumal SR, Srinivasan V, Maestroni GJ, Cardinali DP, Poeggeler B, Hardeland R. Melatonin: Nature's most versatile biological signal? FEBS J 2006; 273(13): 2813-38.
[http://dx.doi.org/10.1111/j.1742-4658.2006.05322.x] [PMID: 16817850]

[85]   Pandi-Perumal SR, Trakht I, Srinivasan V, *et al.* Physiological effects of melatonin: role of melatonin receptors and signal transduction pathways. Prog Neurobiol 2008; 85(3): 335-53.
[http://dx.doi.org/10.1016/j.pneurobio.2008.04.001] [PMID: 18571301]

[86]   Galano A, Tan D, Reiter RJ. Melatonin as a natural ally against oxidative stress: a physicochemical examination 2011; 51: 1-16.
[http://dx.doi.org/10.1111/j.1600-079X.2011.00916.x]

[87]   Hardeland R. Melatonin: signaling mechanisms of a pleiotropic agent. Biofactors 2009; 35(2): 183-92.
[http://dx.doi.org/10.1002/biof.23] [PMID: 19449447]

[88]   Sandyk R, Tsagas N, Anninos PA. Melatonin as a proconvulsive hormone in humans. Int J Neurosci 1992; 63(1-2): 125-35.
[http://dx.doi.org/10.3109/00207459208986662] [PMID: 1342024]

[89]   Banach M, Gurdziel E, Jędrych M, Borowicz KK. Melatonin in experimental seizures and epilepsy. Pharmacol Rep 2011; 63(1): 1-11.
[http://dx.doi.org/10.1016/S1734-1140(11)70393-0] [PMID: 21441606]

[90]   Rocha AK, Cipolla-Neto J, Amado D. Epilepsy: Neuroprotective, anti-inflammatory, and anticonvulsant effects of melatonin.Melatonin: Medical uses and role in health and disease. Nova Science Publishers 2018; pp. 234-53.

[91]   Tchekalarova J, Moyanova S, Fusco AD, Ngomba RT. The role of the melatoninergic system in epilepsy and comorbid psychiatric disorders. Brain Res Bull 2015; 119(Pt A): 80-92.
[http://dx.doi.org/10.1016/j.brainresbull.2015.08.006] [PMID: 26321393]

[92]   Albertson TE, Peterson SL, Stark LG, Lakin ML, Winters WD. The anticonvulsant properties of melatonin on kindled seizures in rats. Neuropharmacology 1981; 20(1): 61-6.
[http://dx.doi.org/10.1016/0028-3908(81)90043-5] [PMID: 7219682]

[93]   Mevissen M, Ebert U. Anticonvulsant effects of melatonin in amygdala-kindled rats. Neurosci Lett 1998; 257(1): 13-6.
[http://dx.doi.org/10.1016/S0304-3940(98)00790-3] [PMID: 9857954]

[94]   Janjoppi L, Silva de Lacerda AF, Scorza FA, Amado D, Cavalheiro EA, Arida RM. Influence of pinealectomy on the amygdala kindling development in rats. Neurosci Lett 2006; 392(1-2): 150-3.
[http://dx.doi.org/10.1016/j.neulet.2005.09.009] [PMID: 16183197]

[95]   Yildirim M, Marangoz C. Anticonvulsant effects of melatonin on penicillin-induced epileptiform activity in rats. Brain Res 2006; 1099(1): 183-8.
[http://dx.doi.org/10.1016/j.brainres.2006.04.093] [PMID: 16764841]

[96]   Aygun H, Aydin D, Inanir S, Ekici F, Ayyildiz M, Agar E. The effects of agomelatine and melatonin on ECoG activity of absence epilepsy model in WAG/Rij rats. Turk J Biol 2015; 39: 904-10.
[http://dx.doi.org/10.3906/biy-1507-32]

[97]   Moyanova S, De Fusco A, Santolini I, *et al.* Abnormal hippocampal melatoninergic system: A potential link between absence epilepsy and depression-like behavior in WAG/Rij Rats? Int J Mol Sci 2018; 19(7): 1973.
[http://dx.doi.org/10.3390/ijms19071973] [PMID: 29986414]

[98]   Aguiar CC, Almeida AB, Araújo PV, *et al.* Anticonvulsant effects of agomelatine in mice. Epilepsy Behav 2012; 24(3): 324-8.

[http://dx.doi.org/10.1016/j.yebeh.2012.04.134] [PMID: 22658946]

[99]   Bikjdaouene L, Escames G, León J, *et al.* Changes in brain amino acids and nitric oxide after melatonin administration in rats with pentylenetetrazole-induced seizures. J Pineal Res 2003; 35(1): 54-60.
       [http://dx.doi.org/10.1034/j.1600-079X.2003.00055.x] [PMID: 12823614]

[100] Dastgheib M, Moezi L. Acute and chronic effects of agomelatine on intravenous penthylenetetrazol-induced seizure in mice and the probable role of nitric oxide. Eur J Pharmacol 2014; 736: 10-5.
       [http://dx.doi.org/10.1016/j.ejphar.2014.04.039] [PMID: 24803306]

[101] Yahyavi-Firouz-Abadi N, Tahsili-Fahadan P, Riazi K, Ghahremani MH, Dehpour AR. Melatonin enhances the anticonvulsant and proconvulsant effects of morphine in mice: role for nitric oxide signaling pathway. Epilepsy Res 2007; 75(2-3): 138-44.
       [http://dx.doi.org/10.1016/j.eplepsyres.2007.05.002] [PMID: 17600683]

[102] Tchekalarova J, Atanasova D, Nenchovska Z, *et al.* Agomelatine protects against neuronal damage without preventing epileptogenesis in the kainate model of temporal lobe epilepsy. Neurobiol Dis 2017; 104: 1-14.
       [http://dx.doi.org/10.1016/j.nbd.2017.04.017] [PMID: 28438504]

[103] Costa-Lotufo LV, Fonteles MM, Lima IS, *et al.* Attenuating effects of melatonin on pilocarpine-induced seizures in rats. Comp Biochem Physiol C Toxicol Pharmacol 2002; 131(4): 521-9.
       [http://dx.doi.org/10.1016/S1532-0456(02)00037-6] [PMID: 11976067]

[104] Lima E, Cabral FR, Cavalheiro EA, Naffah-Mazzacoratti MdaG, Amado D. Melatonin administration after pilocarpine-induced status epilepticus: a new way to prevent or attenuate postlesion epilepsy? Epilepsy Behav 2011; 20(4): 607-12.
       [http://dx.doi.org/10.1016/j.yebeh.2011.01.018] [PMID: 21454134]

[105] Ma Y, Sun X, Li J, *et al.* Melatonin alleviates the epilepsy-associated impairments in hippocampal LTP and spatial learning through rescue of surface GluR2 expression at hippocampal CA1 synapses. Neurochem Res 2017; 42(5): 1438-48.
       [http://dx.doi.org/10.1007/s11064-017-2200-5] [PMID: 28214985]

[106] Rocha AKAA, de Lima E, Amaral F, Peres R, Cipolla-Neto J, Amado D. Altered MT1 and MT2 melatonin receptors expression in the hippocampus of pilocarpine-induced epileptic rats. Epilepsy Behav 2017; 71(Pt A): 23-34.
       [http://dx.doi.org/10.1016/j.yebeh.2017.01.020] [PMID: 28460319]

[107] Atanasova M. Petkova Zl, Pechlivanova D, Dragomirova P, Blazhev, Tchekalarova J. Strain differences in the effect of long-term treatment with melatonin on kainic acid-induced status epilepticus, oxidative stress and the expression of heat shock proteins. Pharmacol Biochem Behav 2013; 111: 44-50.
       [http://dx.doi.org/10.1016/j.pbb.2013.08.006] [PMID: 23978502]

[108] Chung SY, Han SH. Melatonin attenuates kainic acid-induced hippocampal neurodegeneration and oxidative stress through microglial inhibition. J Pineal Res 2003; 34(2): 95-102.
       [http://dx.doi.org/10.1034/j.1600-079X.2003.00010.x] [PMID: 12562500]

[109] Giusti P, Lipartiti M, Franceschini D, Schiavo N, Floreani M, Manev H. Neuroprotection by melatonin from kainate-induced excitotoxicity in rats. FASEB J 1996; 10(8): 891-6.
       [http://dx.doi.org/10.1096/fasebj.10.8.8666166] [PMID: 8666166]

[110] Mohanan PV, Yamamoto HA. Preventive effect of melatonin against brain mitochondria DNA damage, lipid peroxidation and seizures induced by kainic acid. Toxicol Lett 2002; 129(1-2): 99-105.
       [http://dx.doi.org/10.1016/S0378-4274(01)00475-1] [PMID: 11879979]

[111] Petkova Z, Tchekalarova J, Pechlivanova D, *et al.* Treatment with melatonin after status epilepticus attenuates seizure activity and neuronal damage but does not prevent the disturbance in diurnal rhythms and behavioral alterations in spontaneously hypertensive rats in kainate model of temporal lobe epilepsy. Epilepsy Behav 2014; 31: 198-208.

[http://dx.doi.org/10.1016/j.yebeh.2013.12.013] [PMID: 24440891]

[112] Tan D-X, Manchester LC, Reiter RJ, Qi W, Kim SJ, El-Sokkary GH. Melatonin protects hippocampal neurons *in vivo* against kainic acid-induced damage in mice. J Neurosci Res 1998; 54(3): 382-9.
[http://dx.doi.org/10.1002/(SICI)1097-4547(19981101)54:3<382::AID-JNR9>3.0.CO;2-Y] [PMID: 9819143]

[113] Tchekalarova J, Petkova Z, Pechlivanova D, *et al.* Prophylactic treatment with melatonin after status epilepticus: effects on epileptogenesis, neuronal damage, and behavioral changes in a kainate model of temporal lobe epilepsy. Epilepsy Behav 2013; 27(1): 174-87.
[http://dx.doi.org/10.1016/j.yebeh.2013.01.009] [PMID: 23435277]

[114] Uz T, Giusti P, Franceschini D, Kharlamov A, Manev H. Protective effect of melatonin against hippocampal DNA damage induced by intraperitoneal administration of kainate to rats. Neuroscience 1996; 73(3): 631-6.
[http://dx.doi.org/10.1016/0306-4522(96)00155-8] [PMID: 8809783]

[115] Tchekalarova J, Atanasova D, Atanasova M, Kortenska L, Lazarov N. Chronic agomelatine treatment prevents comorbid depression in kainate model of epilepsy through suppression of inflammatory signaling. Neurobiol Dis 2018; 115: 127-44.
[http://dx.doi.org/10.1016/j.nbd.2018.04.005] [PMID: 29653194]

[116] Molina-Carballo A, Muñoz-Hoyos A, Reiter RJ, *et al.* Utility of high doses of melatonin as adjunctive anticonvulsant therapy in a child with severe myoclonic epilepsy: two years' experience. J Pineal Res 1997; 23(2): 97-105.
[http://dx.doi.org/10.1111/j.1600-079X.1997.tb00341.x] [PMID: 9392448]

[117] Guo JF, Yao BZ. [Serum melatonin levels in children with epilepsy or febrile seizures]. Zhongguo Dang Dai Er Ke Za Zhi 2009; 11(4): 288-90.
[PMID: 19374814]

[118] Goldberg-Stern H, Oren H, Peled N, Garty BZ. Effect of melatonin on seizure frequency in intractable epilepsy: a pilot study. J Child Neurol 2012; 27(12): 1524-8.
[http://dx.doi.org/10.1177/0883073811435916] [PMID: 22378657]

[119] Coppola G, Iervolino G, Mastrosimone M, La Torre G, Ruiu F, Pascotto A. Melatonin in wake-sleep disorders in children, adolescents and young adults with mental retardation with or without epilepsy: a double-blind, cross-over, placebo-controlled trial. Brain Dev 2004; 26(6): 373-6.
[http://dx.doi.org/10.1016/j.braindev.2003.09.008] [PMID: 15275698]

[120] Fauteck J, Schmidt H, Lerchl A, Kurlemann G, Wittkowski W. Melatonin in epilepsy: first results of replacement therapy and first clinical results. Biol Signals Recept 1999; 8(1-2): 105-10.
[http://dx.doi.org/10.1159/000014577] [PMID: 10085471]

[121] Gupta M, Aneja S, Kohli K. Add-on melatonin improves sleep behavior in children with epilepsy: randomized, double-blind, placebo-controlled trial. J Child Neurol 2005; 20(2): 112-5.
[http://dx.doi.org/10.1177/08830738050200020501] [PMID: 15794175]

[122] Ijff DM, Aldenkamp AP. Cognitive side-effects of antiepileptic drugs in children. Handb Clin Neurol 2013; 111: 707-18.
[http://dx.doi.org/10.1016/B978-0-444-52891-9.00073-7] [PMID: 23622218]

[123] Uberos J, Augustin-Morales MC, Molina Carballo A, Florido J, Narbona E, Muñoz-Hoyos A. Normalization of the sleep-wake pattern and melatonin and 6-sulphatoxy-melatonin levels after a therapeutic trial with melatonin in children with severe epilepsy. J Pineal Res 2011; 50(2): 192-6.
[PMID: 21044144]

[124] Elkhayat HA, Hassanein SM, Tomoum HY, Abd-Elhamid IA, Asaad T, Elwakkad AS. Melatonin and sleep-related problems in children with intractable epilepsy. Pediatr Neurol 2010; 42(4): 249-54.
[http://dx.doi.org/10.1016/j.pediatrneurol.2009.11.002] [PMID: 20304327]

[125] Gustafsson G, Broström A, Ulander M, Vrethem M, Svanborg E. Occurrence of epileptiform

discharges and sleep during EEG recordings in children after melatonin intake versus sleep-deprivation. Clin Neurophysiol 2015; 126(8): 1493-7.
[http://dx.doi.org/10.1016/j.clinph.2014.10.015] [PMID: 25453612]

[126] Jain SV, Horn PS, Simakajornboon N, *et al.* Melatonin improves sleep in children with epilepsy: a randomized, double-blind, crossover study. Sleep Med 2015; 16(5): 637-44.
[http://dx.doi.org/10.1016/j.sleep.2015.01.005] [PMID: 25862116]

[127] Gupta M, Gupta YK, Agarwal S, Aneja S, Kohli K. A randomized, double-blind, placebo controlled trial of melatonin add-on therapy in epileptic children on valproate monotherapy: effect on glutathione peroxidase and glutathione reductase enzymes. Br J Clin Pharmacol 2004; 58(5): 542-7.
[http://dx.doi.org/10.1111/j.1365-2125.2004.02210.x] [PMID: 15521903]

[128] Gupta M, Gupta YK, Agarwal S, Aneja S, Kalaivani M, Kohli K. Effects of add-on melatonin administration on antioxidant enzymes in children with epilepsy taking carbamazepine monotherapy: a randomized, double-blind, placebo-controlled trial. Epilepsia 2004; 45(12): 1636-9.
[http://dx.doi.org/10.1111/j.0013-9580.2004.17604.x] [PMID: 15571523]

[129] Kopp C, Vogel E, Rettori MC, Delagrange P, Misslin R. The effects of melatonin on the behavioural disturbances induced by chronic mild stress in C3H/He mice. Behav Pharmacol 1999; 10(1): 73-83.
[http://dx.doi.org/10.1097/00008877-199902000-00007] [PMID: 10780304]

[130] Papp M, Gruca P, Boyer PA, Mocaër E. Effect of agomelatine in the chronic mild stress model of depression in the rat. Neuropsychopharmacology 2003; 28(4): 694-703.
[http://dx.doi.org/10.1038/sj.npp.1300091] [PMID: 12655314]

[131] Zhao Y, Fu Y. [Effects of chronic stress depression on the circadian rhythm of peripheral neuroendocrine hormone of rats]. Zhongguo Ying Yong Sheng Li Xue Za Zhi 2017; 33(5): 398-402.
[PMID: 29926582]

[132] Overstreet DH, Pucilowski O, Retton MC, Delagrange P, Guardiola-Lemaitre B. Effects of melatonin receptor ligands on swim test immobility. Neuroreport 1998; 9(2): 249-53.
[http://dx.doi.org/10.1097/00001756-199801260-00014] [PMID: 9507964]

[133] Bertaina-Anglade V, la Rochelle CD, Boyer P-A, Mocaër E. Antidepressant-like effects of agomelatine (S 20098) in the learned helplessness model. Behav Pharmacol 2006; 17(8): 703-13.
[http://dx.doi.org/10.1097/FBP.0b013e3280116e5c] [PMID: 17110796]

[134] Païzanis E, Renoir T, Lelievre V, *et al.* Behavioural and neuroplastic effects of the new-generation antidepressant agomelatine compared to fluoxetine in glucocorticoid receptor-impaired mice. Int J Neuropsychopharmacol 2010; 13(6): 759-74.
[http://dx.doi.org/10.1017/S1461145709990514] [PMID: 19775499]

[135] Morley-Fletcher S, Mairesse J, Soumier A, *et al.* Chronic agomelatine treatment corrects behavioral, cellular, and biochemical abnormalities induced by prenatal stress in rats. Psychopharmacology (Berl) 2011; 217(3): 301-13.
[http://dx.doi.org/10.1007/s00213-011-2280-x] [PMID: 21503609]

[136] Dagytė G, Crescente I, Postema F, *et al.* Agomelatine reverses the decrease in hippocampal cell survival induced by chronic mild stress. Behav Brain Res 2011; 218(1): 121-8.
[http://dx.doi.org/10.1016/j.bbr.2010.11.045] [PMID: 21115070]

[137] Demirdaş A, Nazıroğlu M, Ünal GÖ. Agomelatine reduces brain, kidney and liver oxidative stress but increases plasma cytokine production in the rats with chronic mild stress-induced depression. Metab Brain Dis 2016; 31(6): 1445-53.
[http://dx.doi.org/10.1007/s11011-016-9874-2] [PMID: 27438049]

[138] Redman JR, Francis AJP. Entrainment of rat circadian rhythms by the melatonin agonist S-20098 requires intact suprachiasmatic nuclei but not the pineal. J Biol Rhythms 1998; 13(1): 39-51.
[http://dx.doi.org/10.1177/074873098128999907] [PMID: 9486842]

[139] Højgaard K, Christiansen SL, Bouzinova EV, Wiborg O. Disturbances of diurnal phase markers,

behavior, and clock genes in a rat model of depression; modulatory effects of agomelatine treatment. Psychopharmacology (Berl) 2018; 235(3): 627-40.
[http://dx.doi.org/10.1007/s00213-017-4781-8] [PMID: 29151193]

[140] Sun X, Wang M, Wang Y, *et al.* Melatonin produces a rapid onset and prolonged efficacy in reducing depression-like behaviors in adult rats exposed to chronic unpredictable mild stress. Neurosci Lett 2017; 642: 129-35.
[http://dx.doi.org/10.1016/j.neulet.2017.01.015] [PMID: 28082153]

[141] Mayers AG, Baldwin DS. Antidepressants and their effect on sleep. Hum Psychopharmacol 2005; 20(8): 533-59.
[http://dx.doi.org/10.1002/hup.726] [PMID: 16229049]

[142] Wichniak A, Wierzbicka A, Walęcka M, Jernajczyk W. Effects of antidepressants on sleep. Curr Psychiatry Rep 2017; 19(9): 63.
[http://dx.doi.org/10.1007/s11920-017-0816-4] [PMID: 28791566]

[143] Claustrat B, Chazot G, Brun J, Jordan D, Sassolas G. A chronobiological study of melatonin and cortisol secretion in depressed subjects: plasma melatonin, a biochemical marker in major depression. Biol Psychiatry 1984; 19(8): 1215-28.
[PMID: 6498244]

[144] Rabe-Jabłońska J, Szymańska A. Diurnal profile of melatonin secretion in the acute phase of major depression and in remission. Med Sci Monit 2001; 7(5): 946-52.
[PMID: 11535940]

[145] Thompson C, Franey C, Arendt J, Checkley SA. A comparison of melatonin secretion in depressed patients and normal subjects. Br J Psychiatry 1988; 152: 260-5.
[http://dx.doi.org/10.1192/bjp.152.2.260] [PMID: 3167344]

[146] Leproult R, Van Onderbergen A, L'hermite-Balériaux M, Van Cauter E, Copinschi G. Phase-shifts of 24-h rhythms of hormonal release and body temperature following early evening administration of the melatonin agonist agomelatine in healthy older men. Clin Endocrinol (Oxf) 2005; 63(3): 298-304.
[http://dx.doi.org/10.1111/j.1365-2265.2005.02341.x] [PMID: 16117817]

[147] Sarkisova KY, Midzianovskaia IS, Kulikov MA. Depressive-like behavioral alterations and c-fos expression in the dopaminergic brain regions in WAG/Rij rats with genetic absence epilepsy. Behav Brain Res 2003; 144(1-2): 211-26.
[http://dx.doi.org/10.1016/S0166-4328(03)00090-1] [PMID: 12946611]

[148] Jones NC, Salzberg MR, Kumar G, Couper A, Morris MJ, O'Brien TJ. Elevated anxiety and depressive-like behavior in a rat model of genetic generalized epilepsy suggesting common causation. Exp Neurol 2008; 209(1): 254-60.
[http://dx.doi.org/10.1016/j.expneurol.2007.09.026] [PMID: 18022621]

CHAPTER 5

Modeling Neurodegenerative Diseases Using Transgenic Model of *Drosophila*

Brijesh Singh Chauhan[1]**, Amarish Kumar Yadav**[1]**, Roshan Fatima**[2]**, Sangeeta Arya**[1]**, Jyotsna Singh**[1]**, Rohit Kumar**[1] **and Saripella Srikrishna**[1,*]

[1] *Cell and Neurobiology Laboratory, Department of Biochemistry, Institute of Science, Banaras Hindu University, Varanasi-221005, India*

[2] *National Center for Biological Sciences, Bangalore-560097, India*

Abstract: From the past several decades, neuroscientists have been focusing on understanding the mechanisms of various human neurodegenerative diseases using different models such as *Mouse, Rat, Zebrafish, worm* and *the Drosophila.* Among them, the *Drosophila,* with a short generation time and genetic amenity, has emerged as a vital and prevailing model system to explore multiple aspects of neurodegenerative diseases like Alzheimer's disease, Parkinson's disease, Huntington's disease, Amyotrophic lateral sclerosis, *etc.* In this chapter, we have presented various molecular, genetic and therapeutic approaches employed to model human neurodegenerative diseases using *Drosophila.* Furthermore, we also present the worldwide prevalence of neurodegenerative diseases, along with a survey of published literatures of research conducted in the last two decades on major neurodegenerative diseases employing transgenic *Drosophila,* to evaluate where we stand.

Keywords: Neurodegeneration, Senile plaques, Neurofibrillary tangles, α-Synuclein, Huntingtin, CAG repeat, MARCM system, GAL4 /UAS binary system, CRISPR-Cas system, Therapeutics.

INTRODUCTION

Neurodegenerative disease refers to the gradual loss of neurons of central nervous system (CNS) and peripheral nervous system (PNS) leading to structural and functional damages. The CNS includes brain and spinal cord which control most functions of the body and mind, while PNS includes cranial nerves, peripheral nerves, nerve roots, and neuromuscular junctions positioned outside the brain and spinal cord [1]. Most common neurodegenerative diseases are Alzheimer's disease (AD), Parkinson's disease (PD), Huntington's disease (HD), Frontotem-

*Corresponding author Saripella Srikrishna:Cell and Neurobiology Laboratory; Department of Biochemistry, Institute of Science, Banaras Hindu University, Varanasi.-221005, India; E-mail; skrishna@bhu.ac.in

poral dementia (FTD), Spinocerebellar ataxia (SCA), Multiple sclerosis (MS), and Amyotrophic lateral sclerosis (ALS) [2]. The symptoms of neurodegeneration also manifest in certain other conditions like neuroinfections (due to bacteria and viruses), head trauma, stroke, brain tumors *etc.* [1]. Here, we are focusing on four most prevalent neurodegenerative diseases *i.e.*, AD, PD, HD, and ALS which are quite straight forward for modeling in fruit flies.

In Alzheimer's disease (AD), primarily two candidates, Amyloid-β and Hyper-phosphorylated Tau proteins have been implicated. Over expression/ mutation of the concerned genes lead to neuronal cell death and progressive loss of memory. The amyloidogenic mode of enzymatic action on Amyloid precursor protein (APP) results in Amyloid aggregates over a period of time to form Amyloid plaques. Although, normal function of Amyloid-β is not well understood, plaques evoke numerous neurotoxic effects. On the other hand, hyper-phosphorylated Tau protein leads to formation of neurofibrillary tangles (NFTs). Tau protein is also implicated in the progression of Parkinson's disease, suggesting the susceptibility of AD patients to develop PD symptoms [3]. Several lines of research also indicates greater chances of developing AD like symptom in PD patients and *vice versa* [4, 5]. This might be due to the presence of the common culprit, reactive oxygen species (ROS)/ reactive nitrogen species (RNS), which act as linking agents for neurodegenerative diseases including AD, PD, and HD [6].

Parkinson's disease (PD) is the most common movement disorder. The major proteins involved in PD progression include SNCA (OMIM 163890), Parkin/PARK2 (OMIM 602544), DJ-1 (OMIM 602533) and LRRK2 (OMIM 609007). Mutation and/or misregulation of the genes concerned with these proteins cause neuronal cell death, importantly dopaminergic neurons loss, which ultimately hampers the secretion of dopamine [17].

Huntington's disease (HD), which falls under Polyglutamine (PolyQ) disease group, is a hereditary disease, characterized by progressive loss of brain cells, mainly in ganglion region, and exhibits destruction of mental ability. Previous studies reported that alteration in dopamine (DA) neurotransmission was found in HD patients and also in genetic mouse models of the disease [7, 8]. The modulation in DA transmission level affects the behavioral flexibility and leads to increased risk of Huntington disease [7, 8]. The key protein involved in HD is Huntingtin protein encoded by *HTT* (OMIM 613004) gene. Mutations in this gene lead to growing CAG repeats translated into a PolyQ stretch. The increasing PolyQ stretches manifest in the form of enhancement of motor neuron degeneration. A report revealed that a fifty eight year old male suffering with HD was diagnosed with a coexistence of motor neuron complication, which is an indication of Amyotrophic lateral sclerosis [9].

In ALS, the motor neurons lacking neuronal muscle nourishment cause atrophy or progressive loss of motor neurons affecting the daily work schedule [10]. There are mainly two form of ALS, sporadic and familial. Sporadic ALS is more common and is caused without a clear reason known, accounts upto 90-95% of the cases, while familial ALS shows genetic inheritance and accounts for approximately 5-10% of the cases. However, mutations in genes such as *CHCHD10, TBK1, NEK1, C9orf72* and *SOD1*, enhances the possibility of ALS [11, 12]; In America, familial ALS cases are more prominent due to mutation in genes *c9orf72* (*chromosome 9 open reading frame 72*) and *SOD1 (superoxide dismutase)* [13]. Several reports revealed that mutation in *SOD1* gene causes deposition of misfolded SOD1 proteins in motor neuron. Aggregation of SOD1 proteins, responsible for the mitochondrial dysfunction, ultimately disturb the ATP homeostasis [14]. Therefore, previous studies have been suggesting that these diseases are linked with each other directly or indirectly and often increase the risks of neurodegeneration. Here, we discuss genome wide gene expression studies that link between the neurodegenerative diseases through more genes at a systems wide network level such as, a microglia enriched gene co-expression network is a common link that underlies AD, HD and PD [15]. In past decades, transgenic *Drosophila* has been widely used to understand the molecular aspects of these neurodegenerative-disease progressions, especially, since it offers a simple system for a large scale drug screening.

This book chapter also highlights the key proteins that play an important role in neurodegeneration and the brain regions they affect (Table **1**).

Table 1. Key proteins involved in human-neurodegenerative diseases and the affected brain regions.

S.No.	Disease name	Key proteins	Affected human brain region
1	Alzheimer's disease (AD)	Amyloid-β, Tau	Hippocampus and Cerebral cortex [16]
2	Parkinson's disease (PD)	α-synuclein, Parkin, DJ-1, PINK1(PTEN induced kinase 1), and LRRK2	Basal ganglia and Substantia nigra [17]
3	Huntington's disease (HD)	Huntingtin with expanded polyQ	Basal ganglia [18]
4	Amyotrophic lateral sclerosis (ALS)	SOD1, Ub, p62, FUS, OPTN, TDP-43,ATXN2, UBQLN2, C9ORF72	Motor cortex [19]

TRANSGENIC *DROSOPHILA* TO STUDY NEURODEGENERATIVE DISEASES

Modeling human diseases in animal models and study of the complications involved is a simple approach for understanding the disease mechanisms, where a direct study on human patients is not feasible [20]. There are various animal models available to identify the cause of neuronal diseases. Neuro-researchers centered their attention towards fruit flies due to their small body size, fast generation time, genetic tractability, ease of culture compared to higher model organisms available for study of human diseases [20]. In *Drosophila*, about 2, 50,000 neurons have been reported, against the 86 billions of neurons identified in human brain. Even though there is a huge difference in the number of neurons, they share common features related to neurodegeneration and molecular events [21]. *Drosophila* shows about 75% homology with human disease-causing genes, therefore shares similar functional mechanisms [22, 23]. Major human-neurodegenerative diseases like AD, PD, HD, and ALS have been successfully recapitulated in *Drosophila* [24]. Moreover, genetic screening in transgenic models of *Drosophila* enables us to understand the significance of genes involved in neurodegeneration [20, 25]. There is a possibility for robust genetic manipulation in *Drosophila* to bring about the expression of genes in desired tissues by using specific driver lines or by using neurotoxic chemicals like MPTP (1-methyl-4-phenyl-1,2,3,6-tetrahydropyridine), Paraquat, Rotenone *etc.* to understand the molecular pathogenic events leading to neurodegeneration [26]. Therefore, fruit fly offers a powerful tool for doing molecular studies of human diseases using fly genetics.

Major highlights of this book chapter are models of neurodegenrative diseases like AD, PD, HD and ALS in *Drosophila* and amelioration of these diseases using various therapeutics and through neurodegeneration associated signaling pathways.

ALZHEIMER'S DISEASE (AD) MODELS IN FRUIT FLIES

Alzheimer's disease (AD) is a memory related neurodegenerative disorder first identified in year 1906 by Alois Alzheimer in a 50 year old lady named Auguste Deter. AD is characterized by progressive loss of memory due to the gradual accumulation of Amyloid-β peptide in Hippocampus region and cerebral cortex of the brain [27]. The Amyloid-β peptide is generated through an amyloidogenic pathway involving proteolytic action of β and γ secretases and gradually aggregates to form "senile plaques" in the brain [28, 29]. The other pathological cascade of AD is hyperphosphorylation of Tau protein. The aggregation of hyperphosphorylated Tau proteins in brain cells initiates the formation of paired helical filaments, the accumulation of which results in the formation of neurofibrillary tangles (NFTs) [30].

Drosophila is a well established model organism to study the human AD and the pathogenic proteins of AD, Amyloid-β and Tau, as these are profoundly conserved in human and fly [31]. The Aβ aggregates form Aβ plaques and hyperphosphorylated tau form NFTs are two pathological hallmarks of AD considered as potential drugs targets [32]. These genes have been cloned in *Drosophila* to study various aspects of genetic modulation and for behavioral outcomes to understand the disease progression [33]. RNAi mediated knockdown technology works in *Drosophila* and has revealed signaling pathway constituents [34]. The important signaling pathways like Notch signaling pathway [35], PI3K/Akt pathway [36], JNK/dFOXO pathway [37], and Wnt signaling pathway [35] have been implicated in AD model of *Drosophila*. Therefore, *Drosophila* can be used as a genetic tool for study of genomic alteration, which is possible through various approaches like silencing, deletion, and p-element insertion [31, 38, 39]. On the other hand, behavioral studies in flies have shown that a short-exposure of dim light in night-time disrupted circadian rhythm, which leads to neurodegeneration [40]. For instance, exposure of dim light to tauopathy flies at night for 3days shows circadian rhythm disruption and alteration in sleep-wake cycle due to increase in phosphorylated Tau proteins and neurodegeneration in the AD fly brains [40].

Besides, other biological factors, such as genetic manipulation of metal homeostasis, increases the risks of neurodegeneration in flies. The presence of different metal ions like Copper, Iron, Zinc, and Aluminium influences the Aβ induced neurodegeneration [41, 118]. Hence, in recent years the neuroscientists have explored the role of metals in pathophysiologies of neurodegeration.

In spite of decades of research on the role of biological factors and metals, there are still limited options available for potential cure of AD [42 - 57] (Table **2**).

PARKINSON'S DISEASE (PD) MODELS IN FRUIT FLIES

Parkinson's disease was first described by James Parkinson in the year 1817. PD is a neurodegenerative movement disorder associated with the loss of Dopaminergic (DA) neurons in the mid brain [58]. The motor neuron impairment results in muscle rigidity, slowness of movement, tremors and postural instability [59]. By way of genetic studies several genes namely *α-SNCA, Parkin, Pink, DJ-1etc.* have been shown to be responsible for PD pathogenesis and progression [60]. The use of model organisms like *Drosophila* offers a great opportunity to study multiple aspects of PD development and therapeutics [61]. In *Drosophila,* PD can be modeled either by toxin based induction or by genetic manipulation depending on the context of study [62]. Using UAS-GAL4 system, the over-expression of human PD associated *α-SNCA* gene or its mutant forms A53T or A30P leads to loss of dopaminergic neurons causing locomotor defect, a characteristic of PD [63]. Also, Parkin or PINK gene loss of function results in reduced life span, impairments of motor function, elevated oxidative stress and enhanced dopaminergic neuronal death in fly model of PD. The role of JNK signaling and oxidative stress in PD associated cell death has been explored using fly PD models. Fly PD models have also been utilized to investigate various anti-Parkinson's therapeutics [64 - 76] (Table **2**).

Table 2. List of the most prominent therapeutics that have been used for neurodegenerative diseases in *Drosophila.*

S.No.	Disease name	Therapeutics	Properties
1	Alzheimer's disease (AD)	Flavonoids	Reduce Aβ production [42 - 44]
		Polyphenol; polyhydroxyphenol	Attenuate AD progression [45, 46]
		Calcineurin Inhibitor	Sarah (Nebula), involved in physiological process of memory [47]
		Cl-NQTrp	Alleviates tauopathies [48]
		Coriandrum sativum or Chinese parsley	Suppresses Aβ42 mediated oxidative stress [49]
		Aricept	Traditional Chinese medicine against AD [50]
		Cycloheximide	Degradation of neurofibrillary tangles [51]
		Hydroxylated form of docosahexaenoic acid (DHA-H)	Modifies the AD brain lipid compositions [52]
		Lithium	Suppresses Aβ pathogenesis [53]
		Naphthoquinone-Tryptophan	Reduce tauopathies [54]
		Salidroside	Prevent Aβ and hyperphosphorylation Tau mediated neurotoxicity [36, 55]
		Quercetin	A flavonoid lessen Aβ production [56]
		Metal Chelators	Metal binding proteins implicated in AD [41, 57]
2	Parkinson's disease (PD)	Folic acid	Neuroprotective effect against knockdown of parkin in flies [64]
		Folinic acid	Neuroprotective effect for pink1 in fly model of PD [65]
		Rapamycin	Prevents parkinsonian dopaminergic neuron loss [66]
		Polyphenols	Improve survival and locomotor activity [67]
		Nicotine	Increases lifespan and improve motor activity [68]

(Table 2) cont.....

S.No.	Disease name	Therapeutics	Properties
		Velvet bean	Rescues motor, olfactory, and synaptic impairment [69]
		Curcumin	Neuroprotective efficacy in PD model of *Drosophila* [70]
		Tradinal medicine 'Tianma Gouteng Yin'	Neuroprotective effects in Animal model of PD [71]
		Sevoflurane	Effects on Leucine-rich repeat kinase 2 in PD model of *Drosophila* [72]
		Bromocriptine alginate nanocomposite (BANC)	Neuroprotective effect against PD model of *Drosophila* [73]
		Lovastatin	Rescue neurite degeneration in PD [74]
		Astemizole and Ketoconazole	Neuroprotective effect against PD [75]
		Spirulina	Improves lifespan and locomotor activity in PD [76]
3	Huntington's disease (HD)	AUTEN-67	Hampers the HD progression [94]
		Cystamine and intrabody	Confers neuroprotective advantage in HD model of *Drosophila* [95]
		Camptothecin	Strong suppressor of mutant Hutingtin-induced neurotoxicity [96]
		Mithramycin	Used as chemotherapy for HD [97]
4	Amyotrophic Lateral Sclerosis (ALS)	hnRNP proteins	RNA binding protein for RNA splicing [115]
		Mitofusin/Marf	Ameliorates neuromuscular dysfunction [116]
		Ubiquilin	Protein of ubiquitin family [117]

HUNTINGTON'S DISEASE (HD) MODELS IN FRUIT FLIES

Huntington's disease (HD) is an autosomal dominant neurodegenerative disorder primarily identified by George Huntington in 1892. It is characterized by motor and cognitive abnormality, psychological disturbance and dementia. HD belongs to the CAG/Polyglutamine group of neurodegenerative disorders which also include Spino-cerebellar ataxia (SCAs), fragile X syndrome, spinobulbar muscular atrophy (SBMA), and dentatorubral pallidoluysian atrophy (DRPLA). The main cause of HD is prevalence of more than 35 CAG nucleotide repeats in the *Huntingtin (HTT)* gene [77]. The *HTT* gene plays an important role in early neuronal development but its role in adults is still unknown. However, accumulation of the faulty protein products from CAG repeat enriched *HTT* gene leads to cytotoxicity or neurotoxicity, ultimately resulting in death of the affected neurons. Disease severity and onset are directly correlated with the level of expanded CAG repeat length [78, 79]. However, the exact mechanism of pathophysiology still remains elusive.

The *HTT* gene is present ubiquitously and is conserved from *Drosophila* to human. Human IT15 (Interesting Transcript 15; *HTT* homologue) gene incorporated with different number of polyglutamine (polyQ) repeats such as Q2, Q75 and Q120 have been widely used to model HD using UAS-GAL4 approach in fly [80, 81]. Expression of transgenic human *HTT* gene in *Drosophila* neuronal cells causes disruption of axonal transport and neurotoxicity, which can be directly observed as loss of the ommatidial structure in *Drosophila* using eye specific driver line [82, 83]. In addition, deformity is time dependent so early eclosed flies have less severely distorted eyes than late eclosed flies expressing *HTT* gene, revealing disease progression occurs in an age dependent manner [83].

Previously, various signaling pathways like autophagy, mTOR, SMAD, Kinase, Reactive Oxygen Species (ROS), Calcium, Notch, Wingless/Wnt and BMP, have been reported to be involved in pathogenesis of neurodegenerative disorders including HD *in vitro* [84 - 89]. In SMAD signaling pathway, reduced level of TGF-β (Transforming growth factor-β) in peripheral blood correlates with expanded length of poly glutamine repeat [90]. Another line of evidence concluded that EFGR signaling was altered in glial cells of HD model of *Drosophila* [91]. Brain derived neurotrophic factor (BDNF) is a key protein involved in the neuronal activity and their survival in HD [92]. In addition, HD pathogenesis could possibly be due to the development of oxidative stress [93]. However, its downstream candidates have been poorly explored.

Currently, there are no effective remedies for HD; however various drugs are used as symptomatic treatment. A number of drugs have been successfully screened in transgenic *Drosophila* HD model [94 - 97] (Table **2**).

AMYOTROPHIC LATERAL SCLEROSIS (ALS) MODELS IN FRUIT FLIES

Amyotrophic lateral sclerosis (ALS) or Lou Gehrig disease is a motor neuron related neurodegenerative disorder identified in the year 1869 by Jean-Martin Charcot in a base ball player named Lou Gehrig [98, 99]. In Spite of the past 15 decades of research done on ALS, there has been no cure till date. The disease is characterized by progressive loss of upper and lower motor neurons associated with the voluntary function of neuronal muscles [100]. Some studies reported that mutation in *superoxide dismutase* (*SOD*) gene can lead to ALS [101]. However, the exact cause of ALS is still being debated. Over expression of *SOD* gene augments death of motor neurons by affecting the electrophysiology of neural cells and stress responses in glial cells [102]. This overexpression study demonstrates that *Drosophila* can be used as a tool to investigate and understand the mechanism of pathogenic symptoms and progression of the disease, the signaling pathways involved and therapeutic approaches for its amelioration.

Previous studies have reported that ALS destroys the neuromuscular system of the patient [103]. *Drosophila* model is used as a screening tool to investigate these complications at molecular-genetic level [104 - 107]. A recent study revealed that Superoxide dismutase1 (SOD1) and C9ORF72 dipeptide repeats play a significant role in development of neuromuscular complications [108, 109]. Nevertheless, as exact mechanism of neuromuscular complications in ALS remain elusive, researchers are trying to genetically modify the potential causative agents and to study their effects. For instance, knockdown of *SOD1* gene enables to hold back the downstream severe phenotype effect of ALS [108]. Furthermore, the role of different signaling pathways including BMP [89], Jelly Belly trans-synaptic [110], EGFR [111], Notch [112], JNK [113] and TGF-β [114] have been successfully validated by using ALS model of *Drosophila*. Apart from knockdown, other useful therapeutic approaches have been also well established in transgenic flies [115 - 117] (Table **2**).

MOLECULAR APPROACHES USED FOR MODELING NEURODEGENERATIVE DISEASES IN FRUIT FLIES

The advancement of molecular techniques has opened a great opportunity for fly-researcher to study human diseases in *Drosophila*. Therefore various approaches have been employed to understand mechanisms of neural deterioration in fruit flies. In general, many questions arise such as; why fruit flies are important to study neurodegenerative diseases? What/ How many approaches are available to develop neurodegeneration in fruit flies?

In reply it can be said that, transgenic *Drosophila* are befitting answer to study the function of human proteins involved in neurodegeneration. Studies on neurodegenerative diseases in *Drosophila* model help in understanding the disease mechanisms and their toxic effects [118 - 120]. Thus, to decipher these complications, various molecular approaches have been evolved to generate the symptoms of neurodegeneration in *Drosophila* model.

GAL4/UAS BINARY SYSTEM IN FRUIT FLIES

Researchers globally have adopted tissue-specific GAL4/UAS binary system in *Drosophila* for *in vivo* study of the neurodegenerative diseases [121]. The responder line, Upstream Activating Sequence (UAS) is a very powerful tool used to determine the role of associated genes. The GAL4, an 881 amino acid transcription factor, acquired from yeast *Saccharomyces* turns on the transcription of target gene by binding at cis-regulatory sites of UAS (Fig.1). To study the expression pattern of a preferred human neurodegenerative gene in *Drosophila*, the desired gene is sub-cloned into UAS construct and microinjected in *Drosophila* embryo to create transgenic line. Then, to express the gene, the driver line, GAL4, is used to stimulate the Upstream Activating Sequence (UAS) linked with neurodegenerative genes [20, 122, 123] (Fig.1). Apart from this, other driver lines (QF binary driver and LexA binary driver) are also available to regulate the expression pattern of desired genes [124, 125].

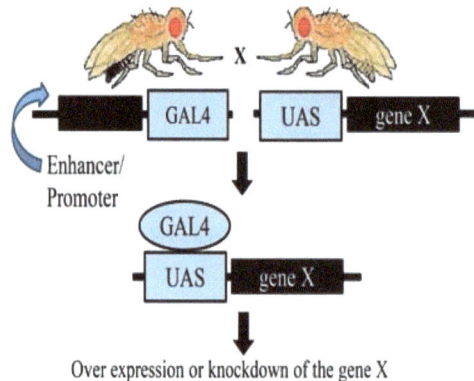

Fig. (1). Schematic diagram representing the GAL4/UAS bipartite system for over expression or knock down of a gene X.

UAS-RNAI (TRIP) APPROACH IN FRUIT FLIES

In *Drosophila* RNA interference (RNAi) approach is widely used for down regulation/silencing of the target gene. Alterations in gene expression changes cell fate and morphology in *Drosophila*. The silencing efficiency for the genes or for any target gene in *Drosophila* tissue depends on the sequence homology of targeted mRNA with RNAi. The knockdown of neurodegenerative genes through RNAi approach is linked to GAL4/UAS system in *Drosophila* [123, 126]. For instance, when two genetically different parental fly lines containing UAS-X[RNAi] and tissue specific driver GAL4 crossed genetically, in their offspring, GAL4 drives the activation of TRiP hairpin transcript under the control of UAS, which binds to the target mRNA of gene X and down regulates its expression (Fig.1).

Initially transgenic RNAi project (TRiP) was developed in the year 2008 at Harvard Medical School and till date over 13,000 fly stocks have been added to TRiP *in vivo* RNAi library for the advantage of the scientific community [127]. These (TRiP) lines are available at flybase, which are induced by GAL4 driver for the specific expression of a hairpin structure to manipulate or alter the expression pattern of desired candidate gene through RNAi [121, 128]. Nowadays, different technologies (i) *in vivo* fly RNAi (TRiP), (ii) *in vivo* CRISPR (TRiP), (iii) cell-based fly RNAi (DRSC) have been developed to knock down the chosen gene [127].

CRISPR/CAS SYSTEM IN FRUIT FLIES

Genome editing tool namely CRISPR-Cas (Clustered Regularly Interspaced Short Palindromic Repeats (CRISPR)-CRISPR-associated protein system (Cas) are used in biological research for genome editing [129]. CRISPR-associated protein 9 nuclease is derived from *Streptococcus pyogenes*. CRISPR-Cas9 editing system has emerged as a robust tool to perform efficient and precise alterations in genomic sequence. It requires a single stranded guide RNA (sgRNA) for targeting the genome sequence for editing. The sgRNA transgenic stock is crossed to a stock in which GAL4 directs expression of a catalytically inactive Cas9 (dCas9) fused to a highly active tripartite activator called VPR (composed of the VP64, P65, and Rta domains) [130]. In the resulting progeny (GAL4>dCas9-VPR; sgRNA-gene) the gene of interest is overexpressed in the GAL4 domain [131]. The schematic representation of gene regulation through CRISPR-Cas9 system in *Drosophila* fly is shown below in (Fig.**2**).

FRUIT FLY MUTANT LINES

Drosophila has emerged as attractive and unique model system. It has significantly contributed to the understanding of age-related human neurodegenerative diseases and has well-developed tools for *in vivo* genetic manipulations. Expression of different disease causing human genes in flies, results in phenotypic changes from overexpression or reduction of endogenous proteins such as Amyloid-β (Aβ), presenilin, microtubule-associated protein Tau (MAPT) in AD; α-synuclein, Parkin, DJ-1 in PD; Polyglutamine (PolyQ) in HD; and SOD1 protein in ALS. Mutations in these human genes hamper the biological functions in *Drosophila* model [81, 102, 132, 133]. Besides these, there are various chemical and environmental factors that cause gene mutation in neurodegenerative disease *Drosophila* model. For instance, Paraquat and rotenone chemically enhance the neurotoxicity in Parkinson's disease [134, 135], imbalance of metal ion like Cu^{2+} induces the Amyloid mediated neurotoxicity [41], environmental factors directly affect the α-synuclein toxicity in PD model of *Drosophila* [25] and mutation in several genes like tau *(MAPT),* progranulin *(GRN)*, transactive response-DNA binding protein-43 *(TARDBP)*, fused in sarcoma *(FUS)* and valosin-containing protein *(VCP)*, enhances possibility of Frontotemporal dementia (FTD) [136].

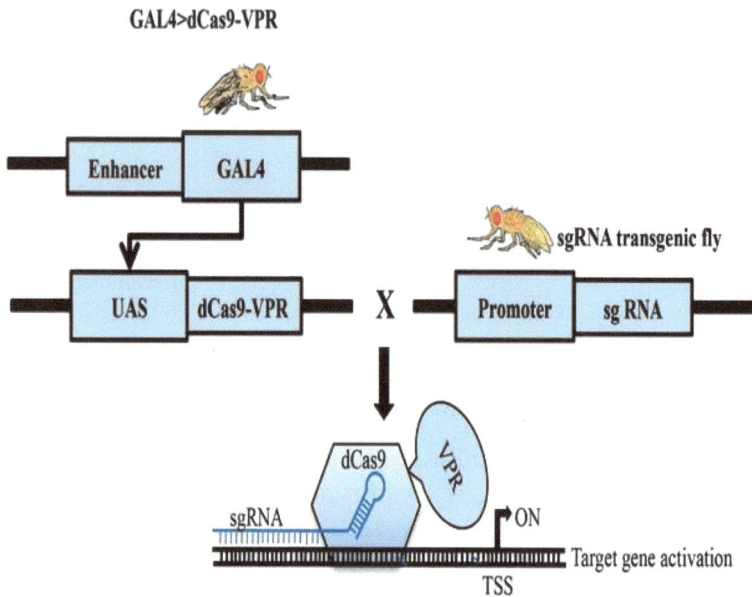

Fig. (2). Schematic representation illustrating the constitutive expression of sgRNAs that target ~500 base pairs upstream of the transcriptional start site (TSS) of a single gene of interest. sgRNA transgenic stock crossed to driver stock which directs expression of a catalytically inactive dead Cas9 (dCas9) fused to a highly active tripartite activator, VPR. In the resulting progeny (GAL4>dCas9-VPR; sgRNA-gene) the gene of interest is over expressed in the GAL4 domain. (Image modified from: https://fgr.hms.harvard.edu/vivo-crispr-0).

MOSAIC ANALYSIS WITH A REPRESSIBLE CELL MARKER (MARCM) SYSTEM

MARCM is a powerful genetic tool widely used in *Drosophila* to label individual homozygous cells against a heterozygous or wild-type cell population. It is composed of UAS-GAL4, GAL80, and Flippase (FLP)/ Flippase recognition target (FRT) for mitotic recombination [137]. Nowadays, MARCM ready transgenic flies are being used to express one or more transgenes. Once "MARCM ready" flies are available, one can generate MARCM clones in just a single cross (MARCM ready cross with FRT mutant fly), as illustrated in Fig.(**3**) and screen flies against balancer. In this technique, Upstream Activating Sequence (UAS) is tagged with reporter genes like GFP or RFP and GAL4 drives expression pattern of the reporter gene in desired cells. TubP-GAL80 acts as a repressor for GAL4 driver lines. For instance, if GAL80 is present in cells, it binds to GAL4 and represses expression of the reporter, thus making cells unlabelled [138, 139]. Study is possible with the combined used of the GAL4/UAS system with the *Saccharomyces cerevisiae* enzyme Flippase recombinase (FLP) that recognizes the 34 bp recombination target sequences (FRTs) on DNA. In the presence of FLP, FRT sites undergo double stranded break and exchange their distal chromosome arms with corresponding FRT sites. The, FLP recombinase is generally tagged with heat shock promoter to control recombination temporally. When cells experience heat shock, expression of FLP recombinase is activated and it recognizes FRT site on DNA, which results in recombination [140]. It brings about the generation of two types of homozygous clones; first one is unlabeled as it has homozygous for GAL80 (a natural suppressor of GAL4). When GAL80 is co-expressed with GAL4, GAL80 inhibits the GAL4 activity. Thus, in presence of GAL80, UAS-Marker making clone unlabeled with GFP. Second clone will have no GAL80, thus GAL4 drives the expression of UAS-Marker making them labeled with GFP. Hence, MARCM system enables the fly researchers to investigate various biological phenomenon including neurodegeneration, neurogenesis and tumor metastasis *etc.* in the vicinity of wild type clone/cells.

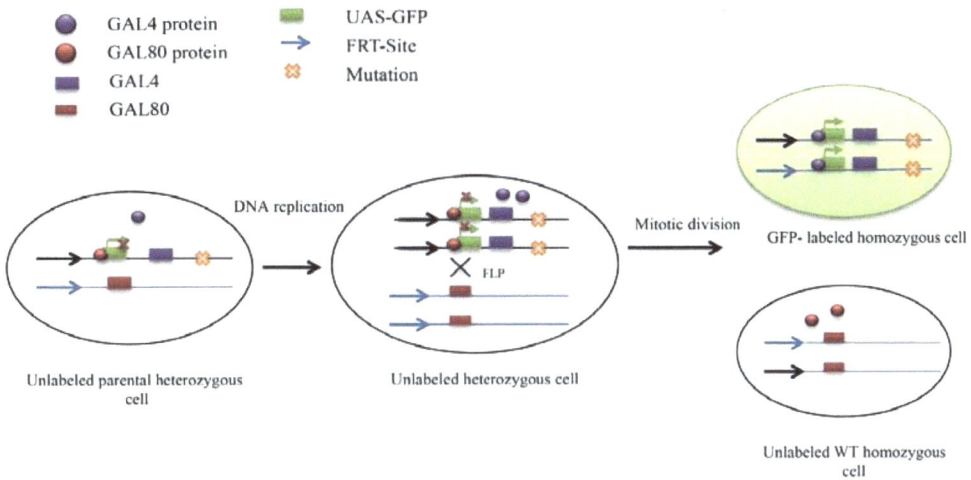

Fig. (3). Schematic digram representing the mosaic analysis with a repressible cell marker. The homozygous mutant clones have no GAL80, therefore GAL4 drives UAS-GFP expression in cell.

WORLDWIDE STATISTICS OF MAJOR NEURODEGENERATIVE DISEASES

Global research statistics of AD, PD, HD and ALS represented here are based on present population and age. Globally, there were more than 21.1 million individuals affected with Alzheimer's dementia in year the 1990 while its number increased and reached upto 46.8 million in 2015. The Alzheimer's dementia cases are expected to increase upto 131.5 million by the year 2050. The prevalence of

AD cases is five times higher in wealthy countries compared to low income countries [141]. Various reports revealed that more than five million Americans are suffering from Alzheimer's dementia [142]. The incidence rate of Alzheimer's dementia in India, specifically South India is greater than North India. It is about 4.86% in the south Indian province of Kerala. By year 2020, cases of Alzheimer's dementia in India are expected to increase up to 14.2% [143].

On the other hand, in case of Parkinson's disease, 6.2 million people are affected globally and around 1, 17400 deaths were reported in between years 1985 and 2010 due to the disease. According to a report [144], more than 100,000 people are living with Parkinson's disease. Certain PD studies have revealed that approximately 41 people of age 40-49, 107 individuals of age 50-59, 428 individuals of the age 60-69, 1,087 individuals of age 70-79 and 1,903 individuals of age 80 and more per 100,000 individuals are globally affected with PD. The worldwide data suggests that the risk of PD is more pronounced in elderly people. The incidence of PD cases is more in males as compared to females and PD prevalences differ according to age variation [145, 146]. The PD prevalences and its possible control in Asian countries are poorly reported.

Worldwide prevalence rate of HD based on meta-analysis is 2.71 per 100,000 people. The incidence of HD in Asian countries is lower (0.40 per 100,000 person) and in Australian, American, and European countries the HD incidence is higher, and it is 5.70 per 100,000 person [147]. In England and Wales, the incidence estimate of HD is 12.4 per 100,000 population in a year [148]. In UK, estimated prevalences of HD are approximately 5.4 per 100,000 person in 1990, 6.4 per 100,000 person in 2004, and 6.6 per 100,000 people in 2007. The HD incidence has been more than two-fold in the year 2010 compared to the prevalence in 1990. Overall HD prevalences in UK are ranges 6-7 per 100,000 people [148, 149]. In Australia, North America, and Europe the incidence rate of HD prevalences was between 15 to 20% [150]. In USA, HD epidemiology was not widespread in year 1993; while it is estimated that approximately 25,000-30,000 individuals have manifested the HD symptoms and further 150,000-250,000 individuals are at risk of developing HD [151, 152].

In case of ALS, the exact estimate of incidence is unknown [153]. However, worldwide incidence of ALS prevalence ranges between 1.5 and 2.5 per 100,000 population in a year [154]. In Europe, occurrence of ALS is almost constant at 2.16 per 100,000 populations in a year [155]. Various reports divulged that frequency of ALS cases is more in males than females in a family. For instance, in population of European origin, incidence of ALS disease was 1:350 for men and 1:400 for women. Sporadic ALS exhibited highest at age of 58-63 and familial ALS at age of 47-52; however ALS incidence decreases rapidly over age of 80 years [155]. In Europe and North America, equal proportion of ALS prevalence occurs, while in Italy, ALS incidence rate lowest 0.6 per 100,000 populations in a year. In contrast, lower incidences are demonstrated among Asia, Africa, and Hispanic ethnicities [153]. Thus, overall statistics reported that ALS incidences are more in American and European people.

Therefore, worldwide probabilities of AD, PD, HD and ALS prevalences will be more in coming future, especially, if precise therapies are not available.

LITERATURE SURVEY FOR NEURODEGENERATIVE DISEASES IN FRUIT FLIES

Abstract of articles published between January 1997 to December 2017 were retrieved from PubMed database (http://www.ncbi.nlm.nih.gov/), using search parameter 'Alzheimer's disease in *Drosophila* and therapeutics' in search box. PD, HD, and ALS diseases were similarly searched in PubMed database. The search was limited to include both review and research articles. Frequency of occurrence of AD, PD, HD and ALS in publications from 1997 to 2007 and 2008 to 2019 is shown as a histogram in (Fig.**4**).

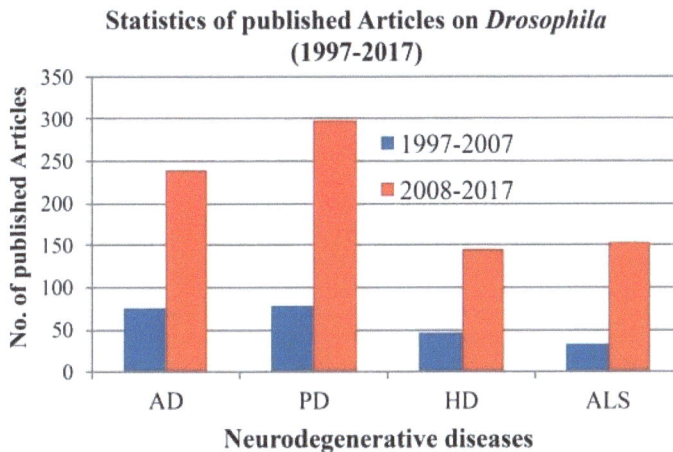

Fig. (4). Histogram represents statistics of literature survey on major neurodegenerative diseases using fruit fly model.

There were total 313 articles on AD, 376 articles on PD, 191 articles on HD, and 186 articles on ALS screened from PubMed in year 1997 to 2017. However, the number of published articles in year 1997-2007 is very less in comparison to articles published in year 2008-2017. Taken together, these results suggest that the past decade (2008-2017) shows increased number of published research articles on AD, HD, PD and ALS compared to in the previous decade (1997 to 2007).

These publication trends show that neurobiologists have increasingly utilized *Drosophila* model for study of neurodegenerative diseases.

CONCLUSIONS AND FUTURE PERSPECTIVES

Nowadays, neurodegenerative complications are a major health concern for human beings. In investigations of the pathophysiology of these diseases, *Drosophila* as a model system has been gaining much attention over the past few decades, and it is being studied using next generation medicinal and molecular biology tools. Although, *Drosophila is* much different from other higher animal model system used to study human neurodegenerative diseases, the disease causing genes show high degree of homology with humans. Therefore, for major neurodegenerative diseases namely AD, PD, HD and ALS more breakthroughs are expected to come from research done using *Drosophila* disease models. Development of advanced molecular techniques including CRISPR in fly offers a great opportunity to further investigate the variety of diseases and discover the specific therapeutic targets.

CONFLICT OF INTEREST:

Authors declare no conflict of interest.

ACKNOWLEDGMENT

This work was supported by a research grant (SERB/EMEQ-389/2014) awarded to S.S. by the Science and Engineering Research Board, India and authors are highly grateful to SERB for financial support.

REFERENCES

[1]   Khan RS, Ahmed MR, Khalid B, Mahmood A, Hassan R. Biomarker Detection of Neurological Disorders through Spectroscopy Analysis. Int Dent Med J Adv Res 2018; 4(1): 1-9.
[http://dx.doi.org/10.15713/ins.idmjar.86]

[2]   Gitler AD, Dhillon P, Shorter J. Neurodegenerative disease: models, mechanisms, and a new hope. Dis Model Mech 2017; 10(5): 499-502.
[http://dx.doi.org/10.1242/dmm.030205] [PMID: 28468935]

[3]   Lei P, Ayton S, Finkelstein DI, Adlard PA, Masters CL, Bush AI. Tau protein: relevance to Parkinson's disease. Int J Biochem Cell Biol 2010; 42(11): 1775-8.
[http://dx.doi.org/10.1016/j.biocel.2010.07.016] [PMID: 20678581]

[4]   Boller F, Mizutani T, Roessmann U, Gambetti P. Parkinson disease, dementia, and Alzheimer disease: clinicopathological correlations. Ann Neurol 1980; 7(4): 329-35.
[http://dx.doi.org/10.1002/ana.410070408] [PMID: 7377758]

[5]   Ditter SM, Mirra SS. Neuropathologic and clinical features of Parkinson's disease in Alzheimer's disease patients. Neurology 1987; 37(5): 754-60.
[http://dx.doi.org/10.1212/WNL.37.5.754] [PMID: 3033544]

[6]   Manoharan S, Guillemin GJ, Abiramasundari RS, Essa MM, Akbar M, Akbar MD. The role of

reactive oxygen species in the pathogenesis of Alzheimer's disease, Parkinson's disease, and Huntington's disease: a mini review. Oxid Med Cell Longev 2016; 20168590578 [http://dx.doi.org/10.1155/2016/8590578] [PMID: 28116038]

[7]   Chen JY, Wang EA, Cepeda C, Levine MS. Dopamine imbalance in Huntington's disease: a mechanism for the lack of behavioral flexibility. Front Neurosci 2013; 7: 114. [http://dx.doi.org/10.3389/fnins.2013.00114] [PMID: 23847463]

[8]   Cepeda C, Murphy KP, Parent M, Levine MS. The role of dopamine in Huntington's disease. Prog Brain Res 2014; 211: 235-54. [http://dx.doi.org/10.1016/B978-0-444-63425-2.00010-6] [PMID: 24968783]

[9]   Phukan J, Ali E, Pender NP, *et al.* Huntington's disease presenting as amyotrophic lateral sclerosis. Amyotroph Lateral Scler 2010; 11(4): 405-7. [http://dx.doi.org/10.3109/17482960903055958] [PMID: 19530012]

[10]   Amyotrophic Lateral Sclerosis Association [homepage on the Internet]. Washington: The ALS Association; [updated: May 2019; cited: 11 August 2019]. 1275 K Street NW, Suite 250-Washington, DC 20005. Available from: http://www.alsa.org

[11]   Brenner D, Müller K, Wieland T, *et al.* NEK1 mutations in familial amyotrophic lateral sclerosis. Brain 2016; 139(Pt 5)e28 [http://dx.doi.org/10.1093/brain/aww033] [PMID: 26945885]

[12]   Amyotrophic Lateral Sclerosis (ALS) Fact Sheet [homepage on the internet]. NIH Bethesda: The ALS Fact Sheet; [updated: 13 August 2019]. Office of Communications and public liaison, National Institute of Neurological Disorders and Stroke, NIH, Bethesda, MD 20892. Available from: https://www.ninds.nih.gov/

[13]   Zarei S, Carr K, Reiley L, *et al.* A comprehensive review of amyotrophic lateral sclerosis. Surg Neurol Int 2015; 6: 171. [http://dx.doi.org/10.4103/2152-7806.169561] [PMID: 26629397]

[14]   Liu J, Lillo C, Jonsson PA, *et al.* Toxicity of familial ALS-linked SOD1 mutants from selective recruitment to spinal mitochondria. Neuron 2004; 43(1): 5-17. [http://dx.doi.org/10.1016/j.neuron.2004.06.016] [PMID: 15233913]

[15]   Mukherjee S, Klaus C, Pricop-Jeckstadt M, Miller JA, Struebing FL. A microglial signature directing human aging and neurodegeneration-related gene networks. Front Neurosci 2019; 13: 2. [http://dx.doi.org/10.3389/fnins.2019.00002] [PMID: 30733664]

[16]   Leinonen V, Koivisto AM, Savolainen S, *et al.* Amyloid and tau proteins in cortical brain biopsy and Alzheimer's disease. Ann Neurol 2010; 68(4): 446-53. [http://dx.doi.org/10.1002/ana.22100] [PMID: 20976765]

[17]   Triarhou LC. Dopamine and Parkinson's disease. In: Madame Curie Bioscience Database [Internet homepage]. Austin (TX): Landes Bioscience; 2000-2013. Available from: https://www.ncbi.nlm.nih.gov/books/NBK6271/

[18]   Reiner A, Dragatsis I, Dietrich P. Genetics and neuropathology of Huntington's disease. Int Rev Neurobiol 2011; 98: 325-72. [http://dx.doi.org/10.1016/B978-0-12-381328-2.00014-6] [PMID: 21907094]

[19]   Agosta F, Valsasina P, Riva N, *et al.* The cortical signature of amyotrophic lateral sclerosis. PLoS One 2012; 7(8)e42816 [http://dx.doi.org/10.1371/journal.pone.0042816] [PMID: 22880116]

[20]   Jackson GR. Guide to understanding *Drosophila* models of neurodegenerative diseases. PLoS Biol 2008; 6(2)e53 [http://dx.doi.org/10.1371/journal.pbio.0060053] [PMID: 18303955]

[21]   Allada R, Chung BY. Circadian organization of behavior and physiology in *Drosophila.* Annu Rev Physiol 2010; 72: 605-24.

[http://dx.doi.org/10.1146/annurev-physiol-021909-135815] [PMID: 20148690]

[22]   Reiter LT, Potocki L, Chien S, Gribskov M, Bier E. A systematic analysis of human disease-associated gene sequences in *Drosophila melanogaster*. Genome Res 2001; 11(6): 1114-25.
[http://dx.doi.org/10.1101/gr.169101] [PMID: 11381037]

[23]   Lloyd TE, Taylor JP. Flightless flies: *Drosophila* models of neuromuscular disease. Ann N Y Acad Sci 2010; 1184: e1-e20.
[http://dx.doi.org/10.1111/j.1749-6632.2010.05432.x] [PMID: 20329357]

[24]   Hirth F. *Drosophila melanogaster* in the study of human neurodegeneration. CNS Neurol Disord-DR Targets (Formerly CNS Neurol Disord Drug Targets) 2010; 9(4): 504-23.
[http://dx.doi.org/10.2174/187152710791556104] [PMID: 20522007]

[25]   Sang TK, Jackson GR. *Drosophila* models of neurodegenerative disease. NeuroRx 2005; 2(3): 438-46.
[http://dx.doi.org/10.1602/neurorx.2.3.438] [PMID: 16389307]

[26]   Şentürk M, Bellen HJ. Genetic strategies to tackle neurological diseases in fruit flies. Curr Opin Neurobiol 2018; 50: 24-32.
[http://dx.doi.org/10.1016/j.conb.2017.10.017] [PMID: 29128849]

[27]   Hippius H, Neundörfer G. The discovery of Alzheimer's disease. Dialogues Clin Neurosci 2003; 5(1): 101-8.
[PMID: 22034141]

[28]   Hui Ping Tan F, Azzam G. *Drosophila melanogaster*: Deciphering Alzheimer's disease. Malaysian J Med Sci 2017; 24(2): 06-20.

[29]   Vivien Chiu WY, Koon AC, Ki Ngo JC, Edwin Chan HY, Lau KF. GULP1/CED-6 ameliorates amyloid-β toxicity in a *Drosophila* model of Alzheimer's disease. Oncotarget 2017; 8(59): 99274-83.
[PMID: 29245900]

[30]   Bolkan BJ, Kretzschmar D. Loss of Tau results in defects in photoreceptor development and progressive neuronal degeneration in *Drosophila*. Dev Neurobiol 2014; 74(12): 1210-25.
[http://dx.doi.org/10.1002/dneu.22199] [PMID: 24909306]

[31]   Moloney A, Sattelle DB, Lomas DA, Crowther DC. Alzheimer's disease: insights from *Drosophila melanogaster* models. Trends Biochem Sci 2010; 35(4): 228-35.
[http://dx.doi.org/10.1016/j.tibs.2009.11.004] [PMID: 20036556]

[32]   Tam C, Wong JH, Ng TB, Tsui SKW, Zuo T. Drugs for targeted therapies of Alzheimer's disease. Curr Med Chem 2019; 26(2): 335-59.
[http://dx.doi.org/10.2174/0929867325666180430150940] [PMID: 29714133]

[33]   Chakraborty R, Vepuri V, Mhatre SD, *et al.* Characterization of a *Drosophila* Alzheimer's disease model: pharmacological rescue of cognitive defects. PLoS One 2011; 6(6)e20799
[http://dx.doi.org/10.1371/journal.pone.0020799] [PMID: 21673973]

[34]   Shulman JM, Feany MB. Genetic modifiers of tauopathy in *Drosophila*. Genetics 2003; 165(3): 1233-42.
[PMID: 14668378]

[35]   Kong Y, Wu J, Zhang D, Wan C, Yuan L. The role of mir-124 in *Drosophila* Alzheimer's disease model by targeting Delta in Notch signaling pathway. Curr Mol Med 2015; 15(10): 980-9.
[http://dx.doi.org/10.2174/1566524016666151123114608] [PMID: 26592243]

[36]   Zhang B, Wang Y, Li H, *et al.* Neuroprotective effects of salidroside through PI3K/Akt pathway activation in Alzheimer's disease models. Drug Des Devel Ther 2016; 10(10): 1335-43.
[PMID: 27103787]

[37]   Hong YK, Lee S, Park SH, *et al.* Inhibition of JNK/dFOXO pathway and caspases rescues neurological impairments in Drosophila Alzheimer's disease model. Biochem Biophys Res Commun 2012; 419(1): 49-53.

[http://dx.doi.org/10.1016/j.bbrc.2012.01.122] [PMID: 22326868]

[38]   Lüchtenborg AM, Katanaev VL. Lack of evidence of the interaction of the Aβ peptide with the Wnt signaling cascade in *Drosophila* models of Alzheimer's disease. Mol Brain 2014; 7(7): 81.
[http://dx.doi.org/10.1186/s13041-014-0081-y] [PMID: 25387847]

[39]   Kennerdell JR, Carthew RW. Heritable gene silencing in *Drosophila* using double-stranded RNA. Nat Biotechnol 2000; 18(8): 896-8.
[http://dx.doi.org/10.1038/78531] [PMID: 10932163]

[40]   Rørth P. A modular misexpression screen in Drosophila detecting tissue-specific phenotypes. Proc Natl Acad Sci USA 1996; 93(22): 12418-22.
[http://dx.doi.org/10.1073/pnas.93.22.12418] [PMID: 8901596]

[41]   Kim M, Subramanian M, Cho YH, Kim GH, Lee E, Park JJ. Short-term exposure to dim light at night disrupts rhythmic behaviors and causes neurodegeneration in fly models of tauopathy and Alzheimer's disease. Biochem Biophys Res Commun 2018; 495(2): 1722-9.
[http://dx.doi.org/10.1016/j.bbrc.2017.12.021] [PMID: 29217196]

[42]   Singh SK, Sinha P, Mishra L, Srikrishna S. Neuroprotective role of a novel copper Chelator against Aβ42 induced neurotoxicity. Int J Alzheimers Dis 2013; 2013567128: 01-10.

[43]   Singh SK, Gaur R, Kumar A, Fatima R, Mishra L, Srikrishna S. The flavonoid derivative 2-(4′ Benzyloxyphenyl)-3-hydroxy-chromen-4-one protects against Aβ42-induced neurodegeneration in transgenic *Drosophila*: insights from in silico and *in vivo* studies. Neurotox Res 2014; 26(4): 331-50.
[http://dx.doi.org/10.1007/s12640-014-9466-z] [PMID: 24706035]

[44]   Baptista FI, Henriques AG, Silva AMS, Wiltfang J, da Cruz e Silva OA. Flavonoids as therapeutic compounds targeting key proteins involved in Alzheimer's disease. ACS Chem Neurosci 2014; 5(2): 83-92.
[http://dx.doi.org/10.1021/cn400213r] [PMID: 24328060]

[45]   Ayoobi F, Shamsizadeh A, Fatemi I, *et al.* Bio-effectiveness of the main flavonoids of Achilleamillefolium in the pathophysiology of neurodegenerative disorders-a review. IJBMS 2017; 20(6): 604-12.

[46]   Jimenez-Del-Rio M, Velez-Pardo C. Alzheimer's disease, *Drosophila melanogaster* and Polyphenols. Adv Exp Med Biol 2015; 863: 21-53.
[http://dx.doi.org/10.1007/978-3-319-18365-7_2] [PMID: 26092625]

[47]   Caruana M, Cauchi R, Vassallo N. Putative role of red wine polyphenols against brain pathology in Alzheimer's and Parkinson's disease. Front Nutr 2016; 3(31): 01-16.
[http://dx.doi.org/10.3389/fnut.2016.00031]

[48]   Lee S, Bang SM, Hong YK, *et al.* The calcineurin inhibitor Sarah (Nebula) exacerbates Aβ42 phenotypes in a *Drosophila* model of Alzheimer's disease. Dis Model Mech 2016; 9(3): 295-306.
[http://dx.doi.org/10.1242/dmm.018069] [PMID: 26659252]

[49]   Frenkel-Pinter M, Tal S, Scherzer-Attali R, *et al.* Cl-NQTrp alleviates tauopathy symptoms in a model organism through the inhibition of tau aggregation-engendered toxicity. Neurodegener Dis 2017; 17(2-3): 73-82.
[http://dx.doi.org/10.1159/000448518] [PMID: 27760426]

[50]   Liu QF, Jeong H, Lee JH, *et al. Coriandrum sativum* suppresses Aβ42-induced ros increases, glial cell proliferation, and ERK activation. Am J Chin Med 2016; 44(7): 1325-47.
[http://dx.doi.org/10.1142/S0192415X16500749] [PMID: 27776428]

[51]   Wang Y, Wang Y, Sui Y, *et al.* The combination of aricept with a traditional Chinese medicine formula, smart soup, may be a novel way to treat Alzheimer's disease. J Alzheimers Dis 2015; 45(4): 1185-95.
[http://dx.doi.org/10.3233/JAD-143183] [PMID: 25690664]

[52]   Geng J, Xia L, Li W, Zhao C, Dou F. Cycloheximide treatment causes a ZVAD-sensitive protease-

dependent cleavage of human tau in *Drosophila* cells. J Alzheimers Dis 2016; 49(4): 1161-8.
[http://dx.doi.org/10.3233/JAD-150423] [PMID: 26599052]

[53]   Mohaibes RJ, Fiol-deRoque MA, Torres M, *et al.* The hydroxylated form of docosahexaenoic acid
       (DHA-H) modifies the brain lipid composition in a model of Alzheimer's disease, improving
       behavioral motor function and survival. Biochim Biophys Acta Biomembr 2017; 1859(9 Pt B): 1596-
       603.
       [http://dx.doi.org/10.1016/j.bbamem.2017.02.020] [PMID: 28284721]

[54]   Sofola-Adesakin O, Castillo-Quan JI, Rallis C, *et al.* Lithium suppresses Aβ pathology by inhibiting
       translation in an adult *Drosophila* model of Alzheimer's disease. Front Aging Neurosci 2014; 6(6):
       190.
       [PMID: 25126078]

[55]   Frenkel-Pinter M, Tal S, Scherzer-Attali R, *et al.* Naphthoquinone tryptophan hybrid inhibits
       aggregation of the Tau derived peptides PHF6 and reduces neurotoxicity. J Alzheimers Dis 2016;
       51(1): 165-78.
       [http://dx.doi.org/10.3233/JAD-150927] [PMID: 26836184]

[56]   Zhang B, Li Q, Chu X, Sun S, Chen S. Salidroside reduces tau hyperphosphorylation *via* up-regulating
       GSK-3β phosphorylation in a tau transgenic *Drosophila* model of Alzheimer's disease. Transl
       Neurodegener 2016; 5: 21.
       [http://dx.doi.org/10.1186/s40035-016-0068-y] [PMID: 27933142]

[57]   Kong Y, Li K, Fu T, *et al.* Quercetin ameliorates Aβ toxicity in *Drosophila* AD model by modulating
       cell cycle-related protein expression. Oncotarget 2016; 7(42): 67716-31.
       [http://dx.doi.org/10.18632/oncotarget.11963] [PMID: 27626494]

[58]   Cristóvão JS, Santos R, Gomes CM. Metals and neuronal metal binding proteins implicated in
       Alzheimer's disease. Oxid Med Cell Longev 2016; 20169812178
       [http://dx.doi.org/10.1155/2016/9812178] [PMID: 26881049]

[59]   Forno LS. Neuropathology of Parkinson's disease. J Neuropathol Exp Neurol 1996; 55(3): 259-72.
       [http://dx.doi.org/10.1097/00005072-199603000-00001] [PMID: 8786384]

[60]   Franco R, Li S, Rodriguez-Rocha H, Burns M, Panayiotidis MI. Molecular mechanisms of pesticide-
       induced neurotoxicity: Relevance to Parkinson's disease. Chem Biol Interact 2010; 188(2): 289-300.
       [http://dx.doi.org/10.1016/j.cbi.2010.06.003] [PMID: 20542017]

[61]   Botella JA, Bayersdorfer F, Gmeiner F, Schneuwly S. Modelling Parkinson's disease in Drosophila.
       Neuromolecular Med 2009; 11(4): 268-80.
       [http://dx.doi.org/10.1007/s12017-009-8098-6] [PMID: 19855946]

[62]   Muñoz-Soriano V, Paricio N. *Drosophila* models of Parkinson's disease: discovering relevant
       pathways and novel therapeutic strategies. Parkinsons Dis 2011; 2011520640
       [http://dx.doi.org/10.4061/2011/520640] [PMID: 21512585]

[63]   Jagmag SA, Tripathi N, Shukla SD, Maiti S, Khurana S. Evaluation of models of Parkinson's disease.
       Front Neurosci 2016; 9: 503.
       [http://dx.doi.org/10.3389/fnins.2015.00503] [PMID: 26834536]

[64]   Feany MB, Pallanck LJ. Parkin: a multipurpose neuroprotective agent? Neuron 2003; 38(1): 13-6.
       [http://dx.doi.org/10.1016/S0896-6273(03)00201-0] [PMID: 12691660]

[65]   Srivastav S, Singh SK, Yadav AK, Srikrishna S. Folic acid supplementation rescues anomalies
       associated with knockdown of parkin in dopaminergic and serotonergic neurons in *Drosophila* model
       of Parkinson's disease. Biochem Biophys Res Commun 2015; 460(3): 780-5.
       [http://dx.doi.org/10.1016/j.bbrc.2015.03.106] [PMID: 25824034]

[66]   Lehmann S, Jardine J, Garrido-Maraver J, Loh SH, Martins LM. Folinic acid is neuroprotective in a
       fly model of Parkinson's disease associated with pink1 mutations. Matters 2017; 3(3)e201702000009
       [http://dx.doi.org/10.19185/matters.201702000009]

[67]   Tain LS, Mortiboys H, Tao RN, Ziviani E, Bandmann O, Whitworth AJ. Rapamycin activation of 4E-BP prevents parkinsonian dopaminergic neuron loss. Nat Neurosci 2009; 12(9): 1129-35.
        [http://dx.doi.org/10.1038/nn.2372] [PMID: 19684592]

[68]   Jimenez-Del-Rio M, Guzman-Martinez C, Velez-Pardo C. The effects of polyphenols on survival and locomotor activity in *Drosophila melanogaster* exposed to iron and paraquat. Neurochem Res 2010; 35(2): 227-38.
        [http://dx.doi.org/10.1007/s11064-009-0046-1] [PMID: 19701790]

[69]   Chambers RP, Call GB, Meyer D, *et al.* Nicotine increases lifespan and rescues olfactory and motor deficits in a *Drosophila* model of Parkinson's disease. Behav Brain Res 2013; 253(253): 95-102.
        [http://dx.doi.org/10.1016/j.bbr.2013.07.020] [PMID: 23871228]

[70]   Poddighe S, De Rose F, Marotta R, *et al.* Mucuna pruriens (Velvet bean) rescues motor, olfactory, mitochondrial and synaptic impairment in PINK1B9 *Drosophila* melanogaster genetic model of Parkinson's disease. PLoS One 2014; 9(10)e110802
        [http://dx.doi.org/10.1371/journal.pone.0110802] [PMID: 25340511]

[71]   Phom L, Achumi B, Alone DP, Muralidhara , Yenisetti SC. Curcumin's neuroprotective efficacy in *Drosophila* model of idiopathic Parkinson's disease is phase specific: implication of its therapeutic effectiveness. Rejuvenation Res 2014; 17(6): 481-9.
        [http://dx.doi.org/10.1089/rej.2014.1591] [PMID: 25238331]

[72]   Liu LF, Song JX, Lu JH, *et al.* Tianma Gouteng Yin, a Traditional Chinese Medicine decoction, exerts neuroprotective effects in animal and cellular models of Parkinson's disease. Sci Rep 2015; 5: 16862.
        [http://dx.doi.org/10.1038/srep16862] [PMID: 26578166]

[73]   Shan Z, Cai S, Zhang T, *et al.* Effects of sevoflurane on leucine-rich repeat kinase 2-associated *Drosophila* model of Parkinson's disease. Mol Med Rep 2015; 11(3): 2062-70.
        [PMID: 25406035]

[74]   Siddique YH, Khan W, Fatima A, *et al.* Effect of bromocriptine alginate nanocomposite (BANC) on a transgenic *Drosophila* model of Parkinson's disease. Dis Model Mech 2016; 9(1): 63-8.
        [http://dx.doi.org/10.1242/dmm.022145] [PMID: 26542705]

[75]   Lin CH, Lin HI, Chen ML, *et al.* Lovastatin protects neurite degeneration in LRRK2-G2019S parkinsonism through activating the Akt/Nrf pathway and inhibiting GSK3β activity. Hum Mol Genet 2016; 25(10): 1965-78.
        [http://dx.doi.org/10.1093/hmg/ddw068] [PMID: 26931464]

[76]   Styczyńska-Soczka K, Zechini L, Zografos L. Validating the Predicted Effect of Astemizole and Ketoconazole Using a *Drosophila* Model of Parkinson's Disease. Assay Drug Dev Technol 2017; 15(3): 106-12.
        [http://dx.doi.org/10.1089/adt.2017.776] [PMID: 28418693]

[77]   Kumar A, Christian PK, Panchal K, Guruprasad BR, Tiwari AK. Supplementation of Spirulina (Arthrospira platensis) Improves Lifespan and Locomotor Activity in Paraquat-Sensitive DJ-1β^{Δ93} Flies, a Parkinson's Disease Model in Drosophila melanogaster. J Diet Suppl 2017; 14(5): 573-88.
        [http://dx.doi.org/10.1080/19390211.2016.1275917] [PMID: 28166438]

[78]   Andrew SE, Goldberg YP, Kremer B, *et al.* The relationship between trinucleotide (CAG) repeat length and clinical features of Huntington's disease. Nat Genet 1993; 4(4): 398-403.
        [http://dx.doi.org/10.1038/ng0893-398] [PMID: 8401589]

[79]   Duyao M, Ambrose C, Myers R, *et al.* Trinucleotide repeat length instability and age of onset in Huntington's disease. Nat Genet 1993; 4(4): 387-92.
        [http://dx.doi.org/10.1038/ng0893-387] [PMID: 8401587]

[80]   Langbehn DR, Hayden MR, Paulsen JS. and the PREDICT-HD Investigators of the Huntington Study Group. CAG-repeat length and the age of onset in Huntington disease (HD): a review and validation study of statistical approaches. Am J Med Genet B Neuropsychiatr Genet 2010; 153B(2): 397-408.

[http://dx.doi.org/10.1002/ajmg.b.30992] [PMID: 19548255]

[81] Jackson GR, Salecker I, Dong X, *et al.* Polyglutamine-expanded human huntingtin transgenes induce degeneration of *Drosophila* photoreceptor neurons. Neuron 1998; 21(3): 633-42.
[http://dx.doi.org/10.1016/S0896-6273(00)80573-5] [PMID: 9768849]

[82] Krench M, Littleton JT. Modeling Huntington disease in *Drosophila*: Insights into axonal transport defects and modifiers of toxicity. Fly (Austin) 2013; 7(4): 229-36.
[http://dx.doi.org/10.4161/fly.26279] [PMID: 24022020]

[83] Gunawardena S, Her LS, Brusch RG, *et al.* Disruption of axonal transport by loss of huntingtin or expression of pathogenic polyQ proteins in *Drosophila.* Neuron 2003; 40(1): 25-40.
[http://dx.doi.org/10.1016/S0896-6273(03)00594-4] [PMID: 14527431]

[84] Kumar JP. Building an ommatidium one cell at a time. Dev Dyn 2012; 241(1): 136-49.
[http://dx.doi.org/10.1002/dvdy.23707] [PMID: 22174084]

[85] Bowles KR, Brooks SP, Dunnett SB, Jones L. Huntingtin subcellular localization is regulated by kinase signaling activity in the StHdhQ111 model of HD. PLoS One 2015; 10(12)e0144864
[http://dx.doi.org/10.1371/journal.pone.0144864] [PMID: 26660732]

[86] Xie W, Wang JQ, Wang QC, Wang Y, Yao S, Tang TS. Adult neural progenitor cells from Huntington's disease mouse brain exhibit increased proliferation and migration due to enhanced calcium and ROS signals. Cell Prolif 2015; 48(5): 517-31.
[http://dx.doi.org/10.1111/cpr.12205] [PMID: 26269226]

[87] Calpena E, delAmo VL, Chakraborty M, *et al.* The Drosophila junctophilin gene is functionally equivalent to its four mammalian counterparts and is a modifier of a Huntingtin poly-Q expansion and the Notch pathway. Dis models Mech 2018; 11(1) dmm029082

[88] Dupont P, Besson MT, Devaux J, Liévens JC. Reducing canonical Wingless/Wnt signaling pathway confers protection against mutant Huntingtin toxicity in *Drosophila.* Neurobiol Dis 2012; 47(2): 237-47.
[http://dx.doi.org/10.1016/j.nbd.2012.04.007] [PMID: 22531500]

[89] Bayat V, Jaiswal M, Bellen HJ. The BMP signaling pathway at the *Drosophila* neuromuscular junction and its links to neurodegenerative diseases. Curr Opin Neurobiol 2011; 21(1): 182-8.
[http://dx.doi.org/10.1016/j.conb.2010.08.014] [PMID: 20832291]

[90] Liévens JC, Rival T, Iché M, Chneiweiss H, Birman S. Expanded polyglutamine peptides disrupt EGF receptor signaling and glutamate transporter expression in *Drosophila.* Hum Mol Genet 2005; 14(5): 713-24.
[http://dx.doi.org/10.1093/hmg/ddi067] [PMID: 15677486]

[91] Zuccato C, Cattaneo E. Role of brain-derived neurotrophic factor in Huntington's disease. Prog Neurobiol 2007; 81(5-6): 294-330.
[http://dx.doi.org/10.1016/j.pneurobio.2007.01.003] [PMID: 17379385]

[92] Browne SE, Beal MF. Oxidative damage in Huntington's disease pathogenesis Prog Neurobiol 2006; 81(5-6): 294-330.
[http://dx.doi.org/10.1089/ars.2006.8.2061]

[93] Battaglia G, Cannella M, Riozzi B, *et al.* Early defect of transforming growth factor β1 formation in Huntington's disease. J Cell Mol Med 2011; 15(3): 555-71.
[http://dx.doi.org/10.1111/j.1582-4934.2010.01011.x] [PMID: 20082658]

[94] Billes V, Kovács T, Hotzi B, *et al.* AUTEN-67 (autophagy enhancer-67) hampers the progression of neurodegenerative symptoms in a *Drosophila* model of Huntington's disease. J Huntingtons Dis 2016; 5(2): 133-47.
[http://dx.doi.org/10.3233/JHD-150180] [PMID: 27163946]

[95] Bortvedt SF, McLear JA, Messer A, Ahern-Rindell AJ, Wolfgang WJ. Cystamine and intrabody co-treatment confers additional benefits in a fly model of Huntington's disease. Neurobiol Dis 2010;

40(1): 130-4.
[http://dx.doi.org/10.1016/j.nbd.2010.04.007] [PMID: 20399860]

[96] Schulte J, Sepp KJ, Wu C, Hong P, Littleton JT. High-content chemical and RNAi screens for suppressors of neurotoxicity in a Huntington's disease model. PLoS One 2011; 6(8)e23841
[http://dx.doi.org/10.1371/journal.pone.0023841] [PMID: 21909362]

[97] Agrawal N, Pallos J, Slepko N, *et al.* Identification of combinatorial drug regimens for treatment of Huntington's disease using *Drosophila.* Proc Natl Acad Sci USA 2005; 102(10): 3777-81.
[http://dx.doi.org/10.1073/pnas.0500055102] [PMID: 15716359]

[98] Gordon PH. Amyotrophic lateral sclerosis: an update for 2013 clinical features, pathophysiology, management and therapeutic trials. Aging Dis 2013; 4(5): 295-310.
[http://dx.doi.org/10.14336/AD.2013.0400295] [PMID: 24124634]

[99] Glicksman MA. The preclinical discovery of amyotrophic lateral sclerosis drugs. Expert Opin Drug Discov 2011; 6(11): 1127-38.
[http://dx.doi.org/10.1517/17460441.2011.628654] [PMID: 22646982]

[100] Wang S, Melhem ER, Poptani H, Woo JH. Neuroimaging in amyotrophic lateral sclerosis. Neurotherapeutics 2011; 8(1): 63-71.
[http://dx.doi.org/10.1007/s13311-010-0011-3] [PMID: 21274686]

[101] Rowland LP, Shneider NA. Amyotrophic lateral sclerosis. N Engl J Med 2001; 344(22): 1688-700.
[http://dx.doi.org/10.1056/NEJM200105313442207] [PMID: 11386269]

[102] Watson MR, Lagow RD, Xu K, Zhang B, Bonini NMA. A drosophila model for amyotrophic lateral sclerosis reveals motor neuron damage by human SOD1. J Biol Chem 2008; 283(36): 24972-81.
[http://dx.doi.org/10.1074/jbc.M804817200] [PMID: 18596033]

[103] Krieger C, Wang SJH, Yoo SH, Harden N. Adducin at the neuromuscular junction in amyotrophic lateral sclerosis: Hanging on for dear life. Front Cell Neurosci 2016; 10: 11.
[http://dx.doi.org/10.3389/fncel.2016.00011] [PMID: 26858605]

[104] Casci I, Pandey UB. A fruitful endeavor: modeling ALS in the fruit fly. Brain Res 2015; 1607: 47-74.
[http://dx.doi.org/10.1016/j.brainres.2014.09.064] [PMID: 25289585]

[105] Bellen HJ, Tong C, Tsuda H. 100 years of *Drosophila* research and its impact on vertebrate neuroscience: a history lesson for the future. Nat Rev Neurosci 2010; 11(7): 514-22.
[http://dx.doi.org/10.1038/nrn2839] [PMID: 20383202]

[106] McGurk L, Berson A, Bonini NM. *Drosophila* as an *in vivo* model for human neurodegenerative disease. Genetics 2015; 201(2): 377-402.
[http://dx.doi.org/10.1534/genetics.115.179457] [PMID: 26447127]

[107] Lu B. Recent advances in using *Drosophila* to model neurodegenerative diseases. Apoptosis 2009; 14(8): 1008-20.
[http://dx.doi.org/10.1007/s10495-009-0347-5] [PMID: 19373559]

[108] Şahin A, Held A, Bredvik K, *et al.* Human SOD1 ALS mutations in a *Drosophila* knock-in model cause severe phenotypes and reveal dosage-sensitive gain-and loss-of-function components. Genetics 2017; 205(2): 707-23.
[http://dx.doi.org/10.1534/genetics.116.190850] [PMID: 27974499]

[109] Lee KH, Zhang P, Kim HJ, *et al.* C9orf72 dipeptide repeats impair the assembly, dynamics, and function of membrane-less organelles. Cell 2016; 167(3): 774-788.e17.
[http://dx.doi.org/10.1016/j.cell.2016.10.002] [PMID: 27768896]

[110] Rohrbough J, Kent KS, Broadie K, Weiss JB. Jelly Belly trans-synaptic signaling to anaplastic lymphoma kinase regulates neurotransmission strength and synapse architecture. Dev Neurobiol 2013; 73(3): 189-208.
[http://dx.doi.org/10.1002/dneu.22056] [PMID: 22949158]

[111] Shimamura M, Kyotani A, Azuma Y, *et al.* Genetic link between Cabeza, a *Drosophila* homologue of Fused in Sarcoma (FUS), and the EGFR signaling pathway. Exp Cell Res 2014; 326(1): 36-45.
[http://dx.doi.org/10.1016/j.yexcr.2014.06.004] [PMID: 24928275]

[112] Yang D, Abdallah A, Li Z, Lu Y, Almeida S, Gao FB. FTD/ALS-associated poly(GR) protein impairs the Notch pathway and is recruited by poly(GA) into cytoplasmic inclusions. Acta Neuropathol 2015; 130(4): 525-35.
[http://dx.doi.org/10.1007/s00401-015-1448-6] [PMID: 26031661]

[113] Zhan L, Xie Q, Tibbetts RS. Opposing roles of p38 and JNK in a *Drosophila* model of TDP-43 proteinopathy reveal oxidative stress and innate immunity as pathogenic components of neurodegeneration. Hum Mol Genet 2015; 24(3): 757-72.
[http://dx.doi.org/10.1093/hmg/ddu493] [PMID: 25281658]

[114] Mushtaq Z, Choudhury SD, Gangwar SK, Orso G, Kumar V. Human senataxin modulates structural plasticity of the neuromuscular junction in *Drosophila* through a neuronally conserved TGFβ signaling pathway. Neurodegener Dis 2016; 16(5-6): 324-36.
[http://dx.doi.org/10.1159/000445435] [PMID: 27197982]

[115] Appocher C, Mohagheghi F, Cappelli S, *et al.* Major hnRNP proteins act as general TDP-43 functional modifiers both in *Drosophila* and human neuronal cells. Nucleic Acids Res 2017; 45(13): 8026-45.
[http://dx.doi.org/10.1093/nar/gkx477] [PMID: 28575377]

[116] Khalil B, Cabirol-Pol MJ, Miguel L, Whitworth AJ, Lecourtois M, Liévens JC. Enhancing Mitofusin/Marf ameliorates neuromuscular dysfunction in *Drosophila* models of TDP-43 proteinopathies. Neurobiol Aging 2017; 54: 71-83.
[http://dx.doi.org/10.1016/j.neurobiolaging.2017.02.016] [PMID: 28324764]

[117] Hanson KA, Kim SH, Wassarman DA, Tibbetts RS. Ubiquilin modifies toxicity of the 43 kilodalton TAR-DNA binding protein (TDP-43) in a *Drosophila* model of amyotrophic lateral sclerosis (ALS). J Biol Chem 2010; 285(15): 11068-72.
[http://dx.doi.org/10.1074/jbc.C109.078527] [PMID: 20154090]

[118] Prüßing K, Voigt A, Schulz JB. *Drosophila melanogaster* as a model organism for Alzheimer's disease. Mol Neurodegener 2013; 8(1): 35.
[http://dx.doi.org/10.1186/1750-1326-8-35] [PMID: 24267573]

[119] Lenz S, Karsten P, Schulz JB, Voigt A. *Drosophila* as a screening tool to study human neurodegenerative diseases. J Neurochem 2013; 127(4): 453-60.
[http://dx.doi.org/10.1111/jnc.12446] [PMID: 24028575]

[120] Chan HY, Bonini NM. *Drosophila* models of human neurodegenerative disease. Cell Death Differ 2000; 7(11): 1075-80.
[http://dx.doi.org/10.1038/sj.cdd.4400757] [PMID: 11139281]

[121] Brand AH, Perrimon N. Targeted gene expression as a means of altering cell fates and generating dominant phenotypes. Development 1993; 118(2): 401-15.
[PMID: 8223268]

[122] Laughon A, Driscoll R, Wills N, Gesteland RF. Identification of two proteins encoded by the Saccharomyces cerevisiae GAL4 gene. Mol Cell Biol 1984; 4(2): 268-75.
[http://dx.doi.org/10.1128/MCB.4.2.268] [PMID: 6366517]

[123] Busson D, Pret AM. GAL4/UAS targeted gene expression for studying Drosophila Hedgehog signaling.In Hedgehog Signaling Protocols Methods Mol Biol. 2007. (pp. 161-201): Humana Press

[124] Stowers RS. An efficient method for recombineering GAL4 and QF drivers. Fly (Austin) 2011; 5(4): 371-8.
[http://dx.doi.org/10.4161/fly.5.4.17560] [PMID: 21857163]

[125] del Valle Rodríguez A, Didiano D, Desplan C. Power tools for gene expression and clonal analysis in *Drosophila.* Nat Methods 2011; 9(1): 47-55.

[http://dx.doi.org/10.1038/nmeth.1800] [PMID: 22205518]

[126]   Schulz JG, David G, Hassan BA. A novel method for tissue-specific RNAi rescue in *Drosophila*. Nucleic Acids Res 2009; 37(13)e93
[http://dx.doi.org/10.1093/nar/gkp450] [PMID: 19483100]

[127]   DRISC/ TRiP Functional Genomics Resources. Available from: https://fgr.hms.harvard.edu/news/new-stocks-added-trip-vivo-rnai-library Jan 6, 2017

[128]   Perkins LA, Holderbaum L, Tao R, *et al*. The transgenic RNAi project at Harvard Medical School: resources and validation. Genetics 2015; 201(3): 843-52.
[http://dx.doi.org/10.1534/genetics.115.180208] [PMID: 26320097]

[129]   Sander JD, Joung JK. CRISPR-Cas systems for editing, regulating and targeting genomes. Nat Biotechnol 2014; 32(4): 347-55.
[http://dx.doi.org/10.1038/nbt.2842] [PMID: 24584096]

[130]   Chavez A, Scheiman J, Vora S, *et al*. Highly efficient Cas9-mediated transcriptional programming. Nat Methods 2015; 12(4): 326-8.
[http://dx.doi.org/10.1038/nmeth.3312] [PMID: 25730490]

[131]   Heidenreich M, Zhang F. Applications of CRISPR-Cas systems in neuroscience. Nat Rev Neurosci 2016; 17(1): 36-44.
[http://dx.doi.org/10.1038/nrn.2015.2] [PMID: 26656253]

[132]   Bertram L, Tanzi RE. The genetic epidemiology of neurodegenerative disease. J Clin Invest 2005; 115(6): 1449-57.
[http://dx.doi.org/10.1172/JCI24761] [PMID: 15931380]

[133]   Marsh JL, Thompson LM. *Drosophila* in the study of neurodegenerative disease. Neuron 2006; 52(1): 169-78.
[http://dx.doi.org/10.1016/j.neuron.2006.09.025] [PMID: 17015234]

[134]   Cassar M, Issa AR, Riemensperger T, *et al*. A dopamine receptor contributes to paraquat-induced neurotoxicity in *Drosophila*. Hum Mol Genet 2015; 24(1): 197-212.
[http://dx.doi.org/10.1093/hmg/ddu430] [PMID: 25158689]

[135]   Bayersdorfer F, Voigt A, Schneuwly S, Botella JA. Dopamine-dependent neurodegeneration in *Drosophila* models of familial and sporadic Parkinson's disease. Neurobiol Dis 2010; 40(1): 113-9.
[http://dx.doi.org/10.1016/j.nbd.2010.02.012] [PMID: 20211259]

[136]   Goedert M, Ghetti B, Spillantini MG. Frontotemporal dementia: implications for understanding Alzheimer disease. Cold Spring Harb Perspect Med 2012; 2(2)a006254
[http://dx.doi.org/10.1101/cshperspect.a006254] [PMID: 22355793]

[137]   Lee T, Luo L. Mosaic analysis with a repressible cell marker for studies of gene function in neuronal morphogenesis. Neuron 1999; 22(3): 451-61.
[http://dx.doi.org/10.1016/S0896-6273(00)80701-1] [PMID: 10197526]

[138]   Lee T, Luo L. Mosaic analysis with a repressible cell marker (MARCM) for *Drosophila* neural development. Trends Neurosci 2001; 24(5): 251-4.
[http://dx.doi.org/10.1016/S0166-2236(00)01791-4] [PMID: 11311363]

[139]   Wu JS, Luo L. A protocol for mosaic analysis with a repressible cell marker (MARCM) in *Drosophila*. Nat Protoc 2006; 1(6): 2583-9.
[http://dx.doi.org/10.1038/nprot.2006.320] [PMID: 17406512]

[140]   Xu T, Rubin GM. Analysis of genetic mosaics in developing and adult *Drosophila* tissues. Development 1993; 117(4): 1223-37.
[PMID: 8404527]

[141]   Prince M, Comas-Herrera A, Knapp M, Guerchet M, Karagiannidou M. World Alzheimer report 2016: improving healthcare for people living with dementia: coverage, quality and costs now and in the

future. Available from: www.alz.co.uk/worldreport 2016

[142] Alzheimer's Association. Alzheimer's disease facts and figures. Alzheimers Dement 2017; 13(4): 325-73.
[http://dx.doi.org/10.1016/j.jalz.2017.02.001]

[143] Mathuranath PS, George A, Ranjith N, *et al.* Incidence of Alzheimer's disease in India: a 10 years follow-up study. Neurol India 2012; 60(6): 625-30.
[http://dx.doi.org/10.4103/0028-3886.105198] [PMID: 23287326]

[144] Parkinson report [homepage on the internet]. New York and Chicago: Parkinson report; [updated: Spring   2017].   Available   from:
http://www.parkinson.org/pd-library/newsletters/Parkinsons-Report-Spring-2017

[145] Pringsheim T, Jette N, Frolkis A, Steeves TD. The prevalence of Parkinson's disease: a systematic review and meta-analysis. Mov Disord 2014; 29(13): 1583-90.
[http://dx.doi.org/10.1002/mds.25945] [PMID: 24976103]

[146] Hirsch L, Jette N, Frolkis A, Steeves T, Pringsheim T. The incidence of Parkinson's disease: a systematic review and meta-analysis. Neuroepidemiology 2016; 46(4): 292-300.
[http://dx.doi.org/10.1159/000445751] [PMID: 27105081]

[147] Pringsheim T, Wiltshire K, Day L, Dykeman J, Steeves T, Jette N. The incidence and prevalence of Huntington's disease: a systematic review and meta-analysis. Mov Disord 2012; 27(9): 1083-91.
[http://dx.doi.org/10.1002/mds.25075] [PMID: 22692795]

[148] Rawlins M. Huntington's disease out of the closet? Lancet 2010; 376(9750): 1372-3.
[http://dx.doi.org/10.1016/S0140-6736(10)60974-9] [PMID: 20594589]

[149] Evans SJ, Douglas I, Rawlins MD, Wexler NS, Tabrizi SJ, Smeeth L. Prevalence of adult Huntington's disease in the UK based on diagnoses recorded in general practice records. J Neurol Neurosurg Psychiatry 2013; 84(10): 1156-60.
[http://dx.doi.org/10.1136/jnnp-2012-304636] [PMID: 23482661]

[150] Rawlins MD, Wexler NS, Wexler AR, *et al.* The prevalence of Huntington's disease. Neuroepidemiology 2016; 46(2): 144-53.
[http://dx.doi.org/10.1159/000443738] [PMID: 26824438]

[151] Frank S. Treatment of Huntington's disease. Neurotherapeutics 2014; 11(1): 153-60.
[http://dx.doi.org/10.1007/s13311-013-0244-z] [PMID: 24366610]

[152] Harper PS. The epidemiology of Huntington's disease. Hum Genet 1992; 89(4): 365-76.
[http://dx.doi.org/10.1007/BF00194305] [PMID: 1535611]

[153] Cronin S, Hardiman O, Traynor BJ. Ethnic variation in the incidence of ALS: a systematic review. Neurology 2007; 68(13): 1002-7.
[http://dx.doi.org/10.1212/01.wnl.0000258551.96893.6f] [PMID: 17389304]

[154] Logroscino G, Traynor BJ, Hardiman O, *et al.* EURALS. Descriptive epidemiology of amyotrophic lateral sclerosis: new evidence and unsolved issues. J Neurol Neurosurg Psychiatry 2008; 79(1): 6-11.
[http://dx.doi.org/10.1136/jnnp.2006.104828] [PMID: 18079297]

[155] Logroscino G, Traynor BJ, Hardiman O, *et al.* EURALS. Incidence of amyotrophic lateral sclerosis in Europe. J Neurol Neurosurg Psychiatry 2010; 81(4): 385-90.
[http://dx.doi.org/10.1136/jnnp.2009.183525] [PMID: 19710046]

CHAPTER 6

Genetic Basis in Stroke Treatment: Targets of Potent Inhibitors

Kanika Vasudeva and **Anjana Munshi**[*]

Department of Human Genetics and Molecular Medicine, Central University of Punjab Bathinda, Punjab, India-151001

Abstract: Stroke is a complex disease resulting from a combination of vascular, environmental and genetic factors. Different therapeutic strategies for the treatment of stroke include antiplatelet therapy, anticoagulants, and lipid- lowering drugs. These drugs act *via* diverse mechanisms of action and target specific enzymes. The enzymes increase the levels of ubiquitous secondary molecules that can cause changes in vascular tone, increase platelet aggregation, cholesterol levels, and other cellular events. Several inhibitors have been developed to curb these enzymes and thus prevent a recurrent stroke. The most potent inhibitors given in the stroke treatment include inhibitors of angiotensin-converting enzyme (ACE) (perindopril, ramipril), phosphodiesterases (PDEs) (rolipram), GpIIb/IIIa and 3-hydroxy-3-methyl-glut-ryl-coenzyme A reductase (HMG-CoA reductase) (pravastatin). ACE inhibitors block the ACE enzyme, thereby preventing the conversion of inactive decapeptide angiotensin I to the active octapeptide and potent vasoconstrictor angiotensin II. Angiotensin II plays a pivotal role in the development of hypertension, atherosclerosis and thrombotic events like stroke. Other inhibitors like phosphodiesterase inhibitors (PDEIs) prevent the inactivation of intracellular mediators of signal transduction such as cAMP and cGMP. These mediators are critical to the regulation of platelet functions. PDEIs are used as antiplatelet agents in clinical settings. Statins are given as lipid-lowering drugs to reduce the risk of stroke by decreasing blood cholesterol levels through inhibition of liver enzyme β-hydroxymethyl glutaryl coenzyme A reductase enzyme. The current chapter will focus on the recent developments in stroke treatment, especially focussing on potent inhibitors such as PDE, ACE, and HMG.

Keywords: Angiotensin-converting enzyme, Compounds, HMG-CoA reductase, Inhibitors, Phosphodiesterase, Platelet activation, Platelet aggregation, SNP, Stroke, Thromboembolism.

---

[*] **Corresponding Author, Prof. Anjana Munshi:** Dean School of Health Sciences, Department of Human Genetics and Molecular Medicine, Central University of Punjab, Bathinda, India; Tel: +919872694373; E-mail: anjanadurani@yahoo.co.in

Atta-ur-Rahman & Zareen Amtul (Eds.)

INTRODUCTION

Stroke is a leading cause of death and neurological disability across the world mainly affecting middle aged and elderly population. However, relatively uncommon young stroke cases are also on the rise, decreasing the average age of ischemic stroke onset. The global stroke burden is increasing worldwide; according to the World Health Organization (WHO) stroke and cerebrovascular diseases kill approximately 5.7 million people each year [1]. The prevalence of stroke is reportedly higher in developing countries, resulting in 75.2% of deaths from stroke and 81.0% of stroke-related disability adjusted life years (DALYs) [2]. Patients who suffer a stroke experience one or more physiological symptoms including weakness, numbness, vision loss, speech difficulties, and motor impairment in terms of facial droop, movements of arm and leg of either or both sides of the body. Immediate diagnosis and treatment are of paramount importance as for every minute in an untreated stroke; 2 million neurons die contributing to significant brain damage [3, 4]. The individuals suspected of having a stroke are assessed by a CT scan or MRI. These scans provide clinicians with a screening tool to decide whether the stroke is ischemic or hemorrhagic. These two stroke subtypes have marked clinical differences in their pathogenesis, prognosis, and treatment [5, 6]. Occlusion in an artery or arteriole of the brain blood supply either by thrombosis or an embolus results in an ischemic stroke while the rupture of blood vessel aneurysm or leakage of a weak vessel in the brain leads to hemorrhagic stroke. In both cases, blood flow in the brain is disrupted causing damage to the downstream brain tissue. Ischemic stroke is a predominant subtype that accounts for the approx. 80% of the total stroke cases while the other 20% suffer from hemorrhagic stroke [7]. Based upon the causes of interruption in blood flow through the perforating arteries of the brain, ischemic stroke can be further classified into several distinct subtypes. The Trial of Org 10172 in Acute Stroke Treatment (TOAST) classified ischemic stroke into 5 main etiological subtypes *i.e.* large artery atherosclerosis, small vessel disease, cardioembolic stroke, stroke of undetermined etiology and other determined etiology [8].

Post-Stroke immediate treatment minimizes long-term disability. Current therapeutic agents for stroke include tissue plasminogen activator, antiplatelet agents, anti-hypertension drugs and lipid-lowering drugs [9]. These drugs act *via* diverse mechanisms of action and are centered on the management of modifiable risk factors like hypertension, diabetes, and hyperlipidemia to prevent recurrent stroke events [10]. Stroke is an immensely complex and personalized disease. Researchers have been focussing on determining the genetic basis of the disease, and a substantial body of evidence suggests that numerous candidate genes are implicated in the pathogenesis of ischemic stroke [11]. Variants of these genes are known to be involved in different pathways like cholesterol biosynthesis, renin-angiotensin aldosterone system (RAS), and cAMP degradation pathway. On account of their crucial molecular role, genes encoding angiotensin-converting enzyme (ACE), phosphodiesterases (PDEs), 3-hydroxy-3-methyl-glut-ryl-coenzyme A reductase (HMG-CoA reductase) and many more have been identified as plausible biological candidates in the development of ischemic stroke [12 - 14]. The variants of these genes have been identified to promote stroke predisposing mediators, *i.e.* enzymes that actively participate in physiological mechanisms [15]. The use of physiological modifiers in molecular therapeutics has emerged as a big avenue in therapeutics. The current chapter will focus on the recent developments in stroke treatment focusing especially on potent inhibitors of PDEs, ACE and HMG Inhibitors.

PHOSPHODIESTERASE (*PDE*)

PDEs are ubiquitously expressed hydrolases. They catalyze the hydrolysis of the 3',5' phosphodiester bond of cAMP and cGMP to yield 5'-AMP and 5'-GMP, respectively [16 - 19]. The regulation of intracellular cyclic nucleotides takes place by 2 enzymes namely adenylyl cyclase (AC) and PDE [20]. Upon their inability to bind to and activate their effectors including protein kinase A, (which is activated by cAMP), protein kinase G (which is activated by cGMP), cyclic-nucleotide-regulated ion channels (which are activated by cAMP or cGMP) and the guanine nucleotide exchange factor Epac (activated by cAMP) [21], vital downstream physiological and pathophysiological processes like cellular growth, differentiation, proliferation, Ca2-dependent signalling, and inflammation, are affected [22]. PDEs constitute a large and complex superfamily that contains 11 PDE gene families (PDE1 to PDE11), comprising 21 genes that generate approximately or more than 100 proteins by alternative splicing and or multiple promoters [23 - 26]. These isoforms are classified based on their cellular function, primary structures, affinities for cAMP and cGMP, catalytic properties, response to specific activators, inhibitors, effectors and their mechanism of regulation. Tissue- specific distribution of PDEs has facilitated targeted pharmacological inhibition in different diseases (Table **1**). They are considered to be key therapeutic targets, from both clinical and economic purposes. Isoforms of PDE on platelets (PDE2, PDE3, PDE5) and brain tissue (PDE1, PDE4, PDE8, PDE9) offer a novel therapeutic strategy in stroke [27]. PDE4D was among the first isoforms of PDEs known to be implicated in the pathogenesis of stroke. 6 SNPs in PDE4D were found to be significantly associated with ischemic stroke in an Icelandic population [28]. Subsequently, several follow up studies in different ethnic groups were taken up to study the SNPs across this gene for association with the disease. A study carried out in the Chinese population also found a significant association of SNP 83 with the disease where carriers of C allele at

SNP 83 had 1.34 times the risk for developing ischemic stroke in comparison to the noncarriers [29]. On account of its underlying role in disease, the mechanism of PDE4D has come up as a major therapeutic target in stroke.

Table 1. Different members of phosphodiesterase isozymes family, their substrates, properties, tissue expression and inhibitors.

S.No.	Family	Substrate	Property	Tissue Expression	Inhibitors	Reference
1	PDE1 (3): PDE1A (nine variants), PDE1B (two variants), PDE1C(five variants	cAMP, cGMP	Ca-calmodulin-stimulated	Heart, brain, lung, smooth muscle	Nimodipline, IC86340, IC22A, dioclein	[17, 30, 31]
2	PDE2 (1): PDE2A (four variants)	cAMP, cGMP	cGMP-stimulated	Adrenal gland, heart,lung,liver,platelets, endothelial cells	EHNA, BAY-60-7750, PDP, IC933, oxindole, ND7001	[17, 30, 31]
3	PDE3 (2): PDE3A (three variants), PDE3B (one variant)	CAMP,cGMP	cAMP-selective, cGMP-inhibited	Heart, smooth muscle, lung, liver, platelets, adipocytes,immunocytes	Cilostamide, milrinone, siguazodan, cilostazol	[17, 30, 31]
4	PDE4 (4): PDE4A (seven variants), PDE4B (four variants), PDE4C (seven variants), PDE4D (nine variants)	cAMP	cAMP-specific, cGMP-insensitive	Brain, sertoli cells, kidney, heart, smooth muscle, lung, liver, endothelial cells, immunocytes	Rolipram, roflumilast, cilomast, NCS 613	[17, 30, 31]
5	PDE5 (1): PDE5A (three variants)	cGMP	cGMP-specific	Lung, platelets,smooth muscle, heart, endothelial cells, brain	Zaprinast, DMPPO, sildenafil, tadalafil, vardenafil	[17, 30, 31]

(Table 1) cont.....

S.No.	Family	Substrate	Property	Tissue Expression	Inhibitors	Reference
6	PDE6 (3): PDE6A (one variant), PDE6B (one variant), PDE6C (one variant)	cGMP	cGMP-specific, Transducin-activated	Photoreceptors, pineal gland, lung	Zaprinast, DMPPO, sildenafil, vardenafil	[17, 30, 31]
7	PDE7 (2): PDE7A (three variants), PDE7B (four variants)	cAMP	cAMP-specific, Rolipram-insensitive	Skeletal muscle, heart, kidney, brain, pancreas, T lymphocytes	BRL 50481, IC242, ASB16165	[17, 30, 31]
8	PDE8 (2): PDE8A (five variants), PDE6B (six variants)	cAMP	cAMP-specific, Rollipram-insensiive IBMX-insensitive	Testes, eye, liver, skeletal musle, heart, kidney, ovary, brain, T lymphocytes, thyroid	PF-04957325	[17, 30, 31]
9	PDE9 (1): PDE9A (twenty variants)	cGMP	cGMP-specific, IBMX-insensitive	Kidney, liver, lung, brain	BAY-73-6691, PF-04447943	[17, 30, 31]
10	PDE10 (1): PDE10A (six variants)	cAMP, cGMP	cGMP-sensitive, cAMP-selective	Testes, brain, thyroid	Papaverine, TP-10, MP-10	[17, 30, 31]
11	PDE11 (1): (four variants)	cAMP, cGMP	cGMP-sensitivity, dual specificity	Skeletal muscle, prostate, pituitary gland, liver, heart	None selective	[17, 30, 31]

(N)= gene numbers.

General Structure of PDEs

The different PDEs share three common structural components 1) a catalytic domain, stretch over of approximately 350 amino acids (aa) folded into 16 helices. Its active site is a deep hydrophobic pocket on which 2 consensus divalent metal-binding domains, *i.e.* Zn^{2+} and Mg^{2+} binding motif (essential for catalytic activity) and a histidine-containing signature motif, $HD(X_2)H(X_4)N$ are located [30, 32]. This enzymatic site encompasses conserved aa residues that form hydrogen bonds with the inhibitors and cyclic nucleotides. Hydrogen bonding results in the formation of 'hydrophobic clamp' with inhibitors that not only anchors them but also orients towards their most active conformation. In addition, an active site also contains variable aa residues that determine affinities and specificities for specific substrates and inhibitors [33]; 2) a regulatory domain between the amino terminus and the catalytic domain. It contains structural determinants that act as GAF (cGMP-specific PDEs, adenylyl cyclases and FhIA binding domain) domain, receptor site for cGMP, for PDE2, PDE5, PDE6, PDE10 and PDE11, PAS domain for PDE8, phosphorylation site for PDE4, autoinhibitory sequence for PDE1 and PDE4 and a membrane association domain for PDE2-4 [21, 34]; 3) a region between the catalytic core and the carboxyl terminus that can be phosphorylated (PDE4) [35] or prenylated (PDE6) by MAPK [36]. The N-terminal of PDE molecules essentially defines the specific properties of each member and variant of the PDE gene family. This region contains targeting domains that are responsible for the localization of PDE isoforms to specific subcellular sites, organelles, membranes and specific signalosomes [37].

Major Therapeutic Inhibitors that have been Identified or Being Developed

PDE inhibitors are potential therapeutic agents for a broad range of diseases namely asthma, inflammation, erectile dysfunction, hypertension, heart failure, thrombosis and cardiac arrhythmias [38]. Previously xanthine derivatives including theophylline and caffeine were used as non-selective PDE inhibitors (PDEIs). Later, PDE5 inhibitor sildenafil got enormous success in treating erectile dysfunction and pulmonary hypertension [39, 40]. In subsequent years, PDEs became a promising therapeutic target for the treatment of many diseases. Despite the therapeutic success of several PDEIs, off-target effects limit their use. At present, only a few PDEIs are widespread in clinical use. In stroke, PDE1, PDE2, PDE3, PDE4, PDE5, PDE8, PDE9 are targeted on account of their expression in brain and platelets [41, 42]. These inhibitors target pathways related to thrombosis, inflammation, platelet aggregation, and neuronal apoptosis. The most extensively used PDEI include vinpoceteine, nimodidipine, cilostazol, rolipram, sildenafil, EHNA, and zaprinast [43, 44].

Members of the PDE1 subfamily have been found to be located mainly in the cytosol where they are activated and regulated by Ca^{2+}calmodulin and the inflow of Ca^{2+} from the extracellular matrix, respectively [45, 46]. PDE1 subfamily comprises the splice variants of three genes namely PDE1A, PDE1B, and PDE1C. These isoforms exhibit unique molecular weights, regulatory properties, Km values, substrate affinities, the association constant for calcium calmodulin, functions and tissue-specific location [47]. Immunostaining studies revealed the localization of PDE1A isoforms in nerve terminals of certain neuron subtypes. PDE1B isoforms are mainly confined to submembrane domains of the cerebellum, hippocampus, and Purkinje cells whereas, PDE1C is limited to neuronal cell body [48]. PDE1 isozymes are critical in regulating vascular smooth muscle contraction, vascular remodeling and neuronal signaling [47]. Inhibition of these isoforms using potential inhibitors prevents the proliferation of smooth muscle cells that subsequently lead to atherosclerosis, one of the main predisposing risk factors in the pathogenesis of stroke [49, 50]. The classical PDE1 inhibitor, vinpocetine, is a derivative of the vinca alkaloid vincamine. It inhibits the PDE1 isoforms with an IC_{50} of approximately 10^{-5} M. Its activity increases the intracellular levels of cAMP and cGMP which leads to the activation of downstream targets protein kinase A and PKG. Vinpocetine is given for the management of vascular diseases including stroke, cerebral hemorrhage, cognitive dysfunctions and improvement of cognition in older individuals [51]. Other non-specific inhibitors of PDE1 include deprenyl, nimodipine, amatadine and caffeine [52]. Patients treated with these inhibitors experience increased cerebral blood flow, improved memory, reduced inflammation, enhanced structural dynamics of dendritic spines and restoration of neuronal plasticity [53, 54].

The inhibition of PDE2 provides a potential therapeutic treatment for the central nervous system (CNS) disorders [55]. A series of purin6-one derivatives showed significant inhibitory potency against PDE2. The two of its best compounds 6p and 6s with IC50 72 and 81 nM, respectively protect HT-22 cells against corticosterone-induced cytotoxicity and rescued corticosterone-induced decreases in cAMP and cGMP levels. PDE2 inhibitors are also known to produce an anxiolytic-like effect in the elevated plus-maze test and exhibited favorable pharmacokinetic properties *in vivo* [56]. PDE3 is an important anti-inflammatory and antiapoptotic signaling mediator. Its isoforms are majorly expressed in platelets [57]. Its inhibition can be achieved by potent inhibitor cilostazol. It induces smooth muscle cell relaxation, platelet inactivation and consequently inhibition of platelet aggregation [58]. On account of its antiplatelet activity, it is recommended to replace conventional antiplatelet therapy by such inhibitors [59]. In a clinical setting, use of PDE3 inhibitor has demonstrated to prevent secondary stroke, recurrent ischemic stroke as well as myocardial infarction [60]. Cilostazol has also reported to reduce angiotensin II-induced abdominal aortic aneurysm formation [61]. Recently, a new PDE-3 inhibitor substance V has proved to have better protective effects in comparison with cilostazol. In a study carried out in the tMCAO model, it was found that PDE-3 inhibition by substance V upregulates cAMP, phospho-PKA, phospho-eNOS, and phospho-AKT, and downregulates NF-κB [62]. Cilostazol treatment also reduces endothelial adhesion molecules and microglia in the brain; therefore, ameliorating vascular cognitive impairment. Altogether it results in enhancement working memory, white matter function and reduced stroke lesions [63, 64]

Rolipram is the first generation inhibitor of PDE4. The other PDE4 inhibitors like Denbufylline (xanthine derivative) and Benzyladenine derivatives were not that effective as they exhibited adverse side effects [65]. Rolipram has anti-inflammatory as well as antithrombotic properties. PDE4D inhibition by rolipram provides neuroprotection against ischemic neurodegeneration. In a study including ischemic stroke patients, it was found that lesser neuronal damage and apoptosis occurred in the rolipram group [66] Fig. (**1**). In addition, rolipram promoted the survival of new neurons by the pharmacological activation of cAMP-CREB signaling which can also provide effective therapy for stroke and post-stroke complications [67]. Besides, rolipram not only enhanced axonal regeneration, but also improved synaptic and cognitive functions after ischemic injury [68]. In an MCAo model of ischemic stroke, it was found that PDE4 inhibitor, FCPR03, protects neuronal cells against apoptosis. It was found to have an inhibitory effect on the generation of reactive oxygen species (ROS) which resulted in the optimum functioning of mitochondria in neuronal cells after cerebral ischemia. FCPR03 also upregulated phosphorylation of Akt and GSK3β and expression of β-catenin in neurons. It altogether protects neurons from brain injury by reducing infarct volume and neurological deficits [69]. Tang and co-workers (2019) recently found that series of novel 3-arylbenzylamine derivatives act as efficient PDE4 inhibitors. The inhibitors with morpholine or pyridin--amine side chain *(e.g.,* 11i, 11l, 11n, 11r–u) displayed good inhibitory activities. Moreover, compound 11r presented improved oral bioavailability and inhibitory activity than its lead compound FCPR03 [70]. CPR16 treatment increased cyclic adenosine monophosphate (cAMP) levels and cAMP-response element binding protein (CREB) phosphorylation in ischemic tissue. In addition, oral administration of 3 mg/kg FCPR16 did not cause vomiting in beagle dogs. This study indicates that FCPR16 has protective effects against cerebral ischemia-reperfusion injury through inhibiting inflammation and apoptosis *via* the cAMP/CREB pathway, while it has low emetogenic potential [71].

Inhibition of PDE5 inhibitor by EMD360527 provides therapeutic benefits in patients with pulmonary arterial hypertension. PDE5 inhibition enhances pulmonary vasodilation and reduction in stroke volume [72]. Sildenafil, another PDE5 inhibitor, reduces the level of hemolysis (LLH) in patients with left ventricular assist device implantation. It significantly reduces the risk of thromboembolic events by decreasing the levels of free plasma hemoglobin, and nitric oxide (NO). This prevents endothelial dysfunction, platelet activation and aggregation [73].

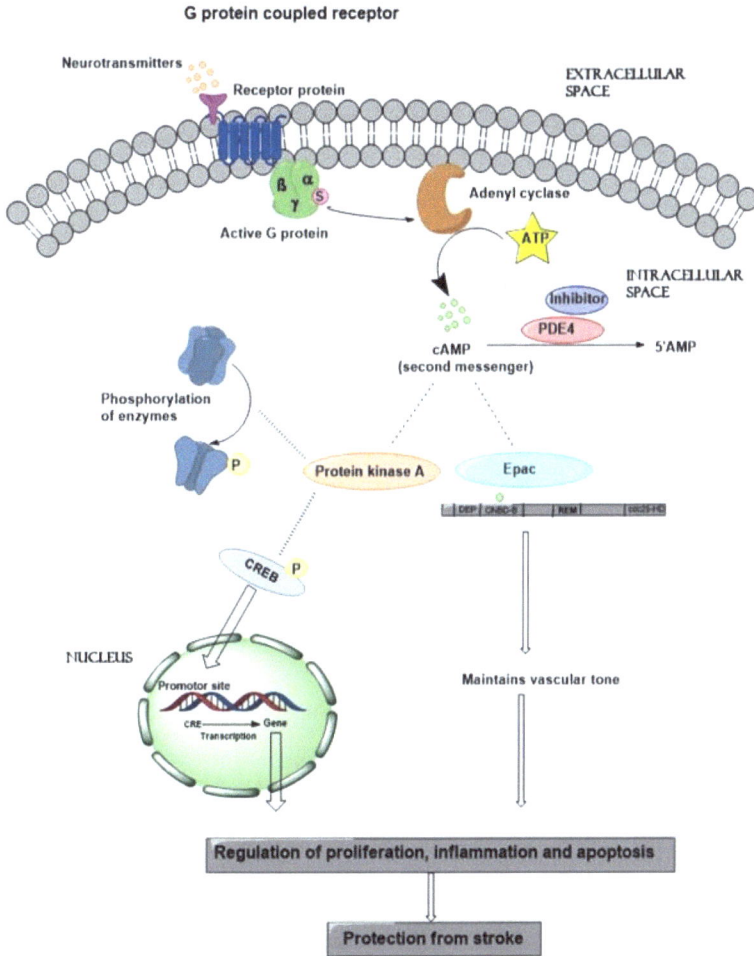

Fig. (1). In the case of cerebral ischemic injury, neurotransmitter molecules through activated G coupled receptor initiate a downstream chain of a cellular reaction. Through adenylyl cyclase (AC) the intracellular secondary messenger, *i.e.* cAMP is generated, which ignites the activation of its downstream effector molecules including protein kinase A, cyclic-nucleotide-regulated ion channels and the guanine nucleotide-exchange factor Epac. Activated PKA phosphorylates proteins like cAMP-responsive element binding protein (CREB) which in turn binds to the cAMP-responsive element and promotes the expression of genes like BDNF. On the other hand, activated EPAC stimulates the phosphorylation of CNBD-b through extracellular signal regulated kinase (1/2) (ERK1/2) signalling. These events promote physiological functions like cellular proliferation, Ca2-dependent signalling, inflammation, and apoptosis. A balance between PDE4 enzyme and adenyl cyclase regulates the level of cAMP. In the ischemic brain, PDE4 activity increases leading to a decrease in the cAMP level. PDE4 enzyme catalyzes the hydrolysis of the 3',5' phosphodiester bond of cAMP to yield 5'-AMP and terminates the cAMP downstream signal transduction. Targeting PDE4 isozymes with inhibitors like Rolipram promotes neuroplasticity in ischemic stroke patients. Inhibition of PDE4 isozymes with inhibitors shows a positive impact on the cAMP/PKA/CREB signalling pathway. As a result, patients exhibit enhanced neuroplasticity, better cognitive functions and improved motor and sensory functions during treatment.

ACE INHIBITORS (ACE-IS)

ACE-Is were first launched in 1970 for treating high blood pressure. They target RAS in plasma as well in the vascular wall. ACEIs inhibit the ACE enzyme, thereby blocking the conversion of angiotensin I to angiotensin II. Various studies including the Heart Outcomes Prevention Evaluation (HOPE) trial and the Perindopril Protection Against Recurrent Stroke Study (PROGRESS) trial have shown that lowering BP with the use of ACE inhibitors (perindopril or ramipril) prevents recurrent stroke [74, 75] (Table 2). Major angiotensin inhibitors available in the market include ramipril, candesartan, and losartan. In HOPE trial, 32% risk reduction of stroke and 20% risk reduction of myocardial infarction (MI) were found among patients who were given ramipril, compared with placebo. The second trial, Losartan Intervention For End-point reduction in hypertension study (LIFE) included 9193 patients with essential hypertension who were randomized to once-daily atenolol or losartan (a selective angiotensin II type 1 receptor blocker); a significant reduction (around 25%) in stroke among patients who were on losartan therapy in comparison with atenolol was observed [76]. The third trial was Study on Cognition and Prognosis in the Elderly (SCOPE) which included 4937 patients with mild hypertension. The patients were treated with candesartan with a dose of 8 mg/per day. It was found to reduce 11% risk of nonfatal stroke, nonfatal MI, or cardiovascular death. The results of these trials suggest that inhibition of generation and action of angiotensin II significantly prevents the risk of vascular diseases including stroke and myocardial infarction [77]. Patients who had a previous stroke or TIA must be treated with ACEIs to lower their BP by means of lifestyle therapy gradually. Recent experiments have demonstrated that treatment with ACE-Is prevents regression and endothelial dysfunction in hypertensive patients. The underlying mechanism involves a decline in cell migration and growth, reduced interstitial fibrosis and improved dysfunction [78]. Another interesting finding is that effects of angiotensin II are achieved by its two receptors, *i.e.* AT1 and AT2. On account of its ability to stimulate AT2 and recruitment of collateral circulation, some studies suggest that AT1 receptor antagonists (AT1RAs) may provide better protection against vascular diseases than ACE-Is. Losartan is a blocker of AT1 receptor. These receptors are also expressed on the surface of platelets. All these factors favour the use of AT1RA in acute stroke for better outcomes [79].

HMG INHIBITORS

The efficacy of statins in the prevention of primary and recurrent ischemic stroke in patients with cardiovascular risks has been well established. A subgroup analysis of the Cholesterol and Recurrent Events Study (CARE) tested the following hypothesis, *i.e.* in patients with average cholesterol, pravastatin

(Pravachol®, Bristol-Myers Squibb, NY, USA) induces aggressive cholesterol reduction and ameliorates the risk of stroke in patients with prior history of myocardial infarction (MI). Of the 4159 subjects included in the study, those who have suffered MI within ten months from the time were enrolled in the study. Compared with placebo, pravastatin reduced total cholesterol by 20% and LDL by 32% and increased high-density lipoprotein (HDL) level by 5% [80 - 85]. Over the five years, there were 52 stroke cases in the treated group versus 76 in the placebo group. When the transient ischemic attacks (TIAs) were included (stroke + TIA), there were 92 events in the pravastatin group versus 124 in the placebo group. The reduction in relative risk (RRR) in the treated group for stroke was 32% and there was a 27% reduction in the combined endpoint of stroke or TIA [81 - 85]. A study evaluating the effectiveness of atorvastatin was conducted in Anglo-Scandinavian Cardiac Outcomes Trial – Lipid Lowering Arm (ASCOT-LLA). This study followed 19,342 hypertensive patients with at least three other cardiovascular risks that were not conventionally determined. Hyperlipidemia in patients was treated with atorvastatin at a dose of 10 mg per day [82]. Consistent follow up of patients was maintained for 5 years. The primary endpoints of the study were nonfatal MI and fatal coronary disease. There was only one primary event in the treated group versus 154 events in the placebo group, which demonstrated a significant reduction of the primary endpoints in the atorvastatin group [83 - 85].

Table 2. Clinical trials including ACEIs in stroke.

S.No	Trials	Cases	Drug	Drug class	Significance	Reference
1.	HOPE	9297	Ramipril	ACEI	32% risk reduction of stroke	[74]
2.	PROGRESS	6105	Perindopril	ACEI	5% decrease in stroke recurrence, 28% decrease in risk reduction	[75]
3.	ONTARGET	25620	Telmisartan/ Ramipril+telmisartan	ACEI+ Angiotensin receptor Blocker	In patients treated with telmisartan, risk of stroke recurrence was reduced	[84]

CONCLUSION

Stroke is one of the leading causes of death worldwide. Effective strategies that prevent stroke recurrence, stroke-related morbidity and mortality are the major concerns of health care agencies worldwide. Prevention of stroke is achieved by the use of antiplatelet drugs, statins, and anticoagulants. Despite the use of these pharmacological drugs, it has been observed that many patients do not respond to the treatment similarly. Researchers have been trying to develop a treatment with maximum drug efficacy and minimize adverse drug reactions. Moreover, recent studies have thus explored the use of PDEI, ACEI and HMG inhibitors in stroke after the successful use of these drugs in a variety of diseases. Several inhibitors namely pravastatin, milrinone, pentoxifylline, and cilostazol are currently in use, each having unique pharmacologic properties. These inhibitors promote a reduction in the active enzymes resulting in inhibition of physiological mediators that lead to stroke. Apart from their therapeutic benefit, these drugs improve functional status, reduces inflammation and oxidative stress. Inhibitors including cilomilast and rolipram have been associated with significant anti-inflammatory effects and reduced side-effects such as nausea, vomiting, and headache.

In conclusion, a devastating neurological emergency like stroke needs to be treated through inhibitors which not only have increased efficacy but also have reduced adverse drug reactions. Both non-selective and selective inhibitors of PDE, ACE, HMG have been used in the treatment of stroke and post-stroke management. Therefore, a deeper understanding of the physiology of ACE, PDEs, and HMG in platelets and other tissues is required for effective antiplatelet therapy. Their full-fledged use in stroke, thus, requires a lot of clinical studies not only concerning their pharmacological properties but also in regards to the adverse side reactions.

CONSENT FOR PUBLICATION

Not applicable.

CONFLICT OF INTEREST

The authors declare that no conflict of interest.

ACKNOWLEDGEMENT

The authors are grateful to the Central University of Punjab, Bathinda, India, for providing the academic, administrative and infrastructural support to carry out this work.

REFERENCES

[1]   Mozaffarian D, Benjamin EJ, Go AS, *et al.* Heart disease and stroke statistics-2016 update a report from the American Heart Association. Circulation 2016; 133(4): e38-e360.
      [PMID: 26673558]

[2]   Feigin VL, Roth GA, Naghavi M, *et al.* Global burden of stroke and risk factors in 188 countries, during 1990-2013: a systematic analysis for the Global Burden of Disease Study 2013. Lancet Neurol 2016; 15(9): 913-24.

[http://dx.doi.org/10.1016/S1474-4422(16)30073-4] [PMID: 27291521]

[3]   Adams HP Jr, Bendixen BH, Kappelle LJ, *et al.* Classification of subtype of acute ischemic stroke. Definitions for use in a multicenter clinical trial. TOAST. Trial of Org 10172 in Acute Stroke Treatment. Stroke 1993; 24(1): 35-41.
[http://dx.doi.org/10.1161/01.STR.24.1.35] [PMID: 7678184]

[4]   Newton AJH, Lytton WW. Computer modeling of ischemic stroke. Drug Discov Today Dis Models 2016; 19: 77-83.
[http://dx.doi.org/10.1016/j.ddmod.2017.01.001] [PMID: 28943884]

[5]   Jovin TG, Chamorro A, Cobo E, *et al.* Thrombectomy within 8 hours after symptom onset in ischemic stroke. N Engl J Med 2015; 372(24): 2296-306.
[http://dx.doi.org/10.1056/NEJMoa1503780] [PMID: 25882510]

[6]   Zille M, Karuppagounder SS, Chen Y, *et al.* Neuronal death after hemorrhagic stroke *in vitro* and *in vivo* shares features of ferroptosis and necroptosis. Stroke 2017; 48(4): 1033-43.
[http://dx.doi.org/10.1161/STROKEAHA.116.015609] [PMID: 28250197]

[7]   Feigin VL, Krishnamurthi RV, Parmar P, Norrving B, Mensah GA, Bennett DA, *et al.* Update on the global burden of ischemic and hemorrhagic stroke in 1990-2013: The GBD 2013 study 2015.

[8]   Dziadkowiak E, Chojdak-Łukasiewicz J, Guziński M, Noga L, Paradowski B. The usefulness of the TOAST classification and prognostic significance of pyramidal symptoms during the acute phase of cerebellar ischemic stroke. Cerebellum 2016; 15(2): 159-64.
[http://dx.doi.org/10.1007/s12311-015-0676-6] [PMID: 26041073]

[9]   Vasudeva K, Chaurasia P, Singh S, Munshi A. Genetic signatures in ischemic stroke: Focus on aspirin resistance CNS & Neurological Disorders-Drug Targets (Formerly Current Drug Targets-CNS & Neurological Disorders) 2017; 16(9): 974-82.

[10]   Mach F, Ray KK, Wiklund O, *et al.* Adverse effects of statin therapy: perception vs. the evidence - focus on glucose homeostasis, cognitive, renal and hepatic function, haemorrhagic stroke and cataract. Eur Heart J 2018; 39(27): 2526-39.
[http://dx.doi.org/10.1093/eurheartj/ehy182] [PMID: 29718253]

[11]   Munshi A, Das S, Kaul S. Genetic determinants in ischaemic stroke subtypes: seven year findings and a review. Gene 2015; 555(2): 250-9.
[http://dx.doi.org/10.1016/j.gene.2014.11.015] [PMID: 25447900]

[12]   Das S, Roy S, Sharma V, Kaul S, Jyothy A, Munshi A. Association of ACE gene I/D polymorphism and ACE levels with hemorrhagic stroke: comparison with ischemic stroke. Neurol Sci 2015; 36(1): 137-42.
[http://dx.doi.org/10.1007/s10072-014-1880-8] [PMID: 25015258]

[13]   Munshi A, Das S. Genetic understanding of stroke treatment: potential role for phosphodiesterase inhibitors Phosphodiesterases: CNS Functions and Diseases. Springer 2017; pp. 445-61.
[http://dx.doi.org/10.1007/978-3-319-58811-7_16]

[14]   Das S, Kaul S, Jyothy A, Munshi A. Role of TLR4 (C1196T) and CD14 (C-260T) Polymorphisms in Development of Ischemic Stroke, Its Subtypes and Hemorrhagic Stroke. J Mol Neurosci 2017; 63(3-4): 300-7.
[http://dx.doi.org/10.1007/s12031-017-0979-9] [PMID: 28963650]

[15]   Malik R, Traylor M, Pulit SL, *et al.* Low-frequency and common genetic variation in ischemic stroke: The METASTROKE collaboration. Neurology 2016; 86(13): 1217-26.
[http://dx.doi.org/10.1212/WNL.0000000000002528] [PMID: 26935894]

[16]   Azevedo MF, Faucz FR, Bimpaki E, *et al.* Clinical and molecular genetics of the phosphodiesterases (PDEs). Endocr Rev 2014; 35(2): 195-233.
[http://dx.doi.org/10.1210/er.2013-1053] [PMID: 24311737]

[17]   Lugnier C. Cyclic nucleotide phosphodiesterase (PDE) superfamily: a new target for the development

of specific therapeutic agents. Pharmacol Ther 2006; 109(3): 366-98.
[http://dx.doi.org/10.1016/j.pharmthera.2005.07.003] [PMID: 16102838]

[18]   Fajardo AM, Piazza GA, Tinsley HN. The role of cyclic nucleotide signaling pathways in cancer: targets for prevention and treatment. Cancers (Basel) 2014; 6(1): 436-58.
[http://dx.doi.org/10.3390/cancers6010436] [PMID: 24577242]

[19]   Card GL, Blasdel L, England BP, *et al.* A family of phosphodiesterase inhibitors discovered by cocrystallography and scaffold-based drug design. Nat Biotechnol 2005; 23(2): 201-7.
[http://dx.doi.org/10.1038/nbt1059] [PMID: 15685167]

[20]   Bender AT, Beavo JA. Cyclic nucleotide phosphodiesterases: molecular regulation to clinical use. Pharmacol Rev 2006; 58(3): 488-520.
[http://dx.doi.org/10.1124/pr.58.3.5] [PMID: 16968949]

[21]   Houslay MD, Baillie GS, Maurice DH. cAMP-Specific phosphodiesterase-4 enzymes in the cardiovascular system: a molecular toolbox for generating compartmentalized cAMP signaling. Circ Res 2007; 100(7): 950-66.
[http://dx.doi.org/10.1161/01.RES.0000261934.56938.38] [PMID: 17431197]

[22]   Thompson WJ, Terasaki WL, Epstein PM, Strada SJ. Assay of cyclic nucleotide phosphodiesterase and resolution of multiple molecular forms of the enzyme. Adv Cyclic Nucleotide Res 1979; 10: 69-92.
[PMID: 222125]

[23]   Thompson WJ, Appleman MM. Cyclic nucleotide phosphodiesterase and cyclic AMP. Ann N Y Acad Sci 1971; 185(1): 36-41.
[http://dx.doi.org/10.1111/j.1749-6632.1971.tb45233.x] [PMID: 4330503]

[24]   Soderling SH, Bayuga SJ, Beavo JA. Cloning and characterization of a cAMP-specific cyclic nucleotide phosphodiesterase. Proc Natl Acad Sci USA 1998; 95(15): 8991-6.
[http://dx.doi.org/10.1073/pnas.95.15.8991] [PMID: 9671792]

[25]   Conti M, Beavo J. Biochemistry and physiology of cyclic nucleotide phosphodiesterases: essential components in cyclic nucleotide signaling. Annu Rev Biochem 2007; 76: 481-511.
[http://dx.doi.org/10.1146/annurev.biochem.76.060305.150444] [PMID: 17376027]

[26]   Essayan DM. Cyclic nucleotide phosphodiesterases. J Allergy Clin Immunol 2001; 108(5): 671-80.
[http://dx.doi.org/10.1067/mai.2001.119555] [PMID: 11692087]

[27]   Card GL, England BP, Suzuki Y, *et al.* Structural basis for the activity of drugs that inhibit phosphodiesterases. Structure 2004; 12(12): 2233-47.
[http://dx.doi.org/10.1016/j.str.2004.10.004] [PMID: 15576036]

[28]   Gretarsdottir S, Thorleifsson G, Reynisdottir ST, *et al.* The gene encoding phosphodiesterase 4D confers risk of ischemic stroke. Nat Genet 2003; 35(2): 131-8.
[http://dx.doi.org/10.1038/ng1245] [PMID: 14517540]

[29]   Xu X, Li X, Li J, Ou R, Sheng W. Meta-analysis of association between variation in the PDE4D gene and ischemic cerebral infarction risk in Asian populations. Neurogenetics 2010; 11(3): 327-33.
[http://dx.doi.org/10.1007/s10048-010-0235-8] [PMID: 20182758]

[30]   Maurice DH, Ke H, Ahmad F, Wang Y, Chung J, Manganiello VC. Advances in targeting cyclic nucleotide phosphodiesterases. Nat Rev Drug Discov 2014; 13(4): 290-314.
[http://dx.doi.org/10.1038/nrd4228] [PMID: 24687066]

[31]   Keravis T, Lugnier C. Cyclic nucleotide phosphodiesterase (PDE) isozymes as targets of the intracellular signalling network: benefits of PDE inhibitors in various diseases and perspectives for future therapeutic developments. Br J Pharmacol 2012; 165(5): 1288-305.
[http://dx.doi.org/10.1111/j.1476-5381.2011.01729.x] [PMID: 22014080]

[32]   Rehmann H, Wittinghofer A, Bos JL. Capturing cyclic nucleotides in action: snapshots from crystallographic studies. Nat Rev Mol Cell Biol 2007; 8(1): 63-73.

[http://dx.doi.org/10.1038/nrm2082] [PMID: 17183361]

[33]   Lee LC, Maurice DH, Baillie GS. Targeting protein-protein interactions within the cyclic AMP signaling system as a therapeutic strategy for cardiovascular disease. Future Med Chem 2013; 5(4): 451-64.
       [http://dx.doi.org/10.4155/fmc.12.216] [PMID: 23495691]

[34]   Manganiello VC, Degerman E. Cyclic nucleotide phosphodiesterases (PDEs): diverse regulators of cyclic nucleotide signals and inviting molecular targets for novel therapeutic agents. Thromb Haemost 1999; 82(2): 407-11.
       [PMID: 10605731]

[35]   Francis SH, Turko IV, Corbin JD. Cyclic nucleotide phosphodiesterases: relating structure and function 2000.
       [http://dx.doi.org/10.1016/S0079-6603(00)65001-8]

[36]   Cote RH. Characteristics of photoreceptor PDE (PDE6): similarities and differences to PDE5. Int J Impot Res 2004; 16(S1) (Suppl. 1): S28-33.
       [http://dx.doi.org/10.1038/sj.ijir.3901212] [PMID: 15224133]

[37]   Essayan DM. Cyclic nucleotide phosphodiesterase (PDE) inhibitors and immunomodulation. Biochem Pharmacol 1999; 57(9): 965-73.
       [http://dx.doi.org/10.1016/S0006-2952(98)00331-1] [PMID: 10796066]

[38]   Ke H, Wang H. Crystal structures of phosphodiesterases and implications on substrate specificity and inhibitor selectivity. Curr Top Med Chem 2007; 7(4): 391-403.
       [http://dx.doi.org/10.2174/156802607779941242] [PMID: 17305581]

[39]   Ghofrani HA, Osterloh IH, Grimminger F. Sildenafil: from angina to erectile dysfunction to pulmonary hypertension and beyond. Nat Rev Drug Discov 2006; 5(8): 689-702.
       [http://dx.doi.org/10.1038/nrd2030] [PMID: 16883306]

[40]   Sandner P, Hütter J, Tinel H, Ziegelbauer K, Bischoff E. PDE5 inhibitors beyond erectile dysfunction. Int J Impot Res 2007; 19(6): 533-43.
       [http://dx.doi.org/10.1038/sj.ijir.3901577] [PMID: 17625575]

[41]   Haslam RJ, Dickinson NT, Jang EK. Cyclic nucleotides and phosphodiesterases in platelets. Thromb Haemost 1999; 82(2): 412-23.
       [PMID: 10605732]

[42]   Lakics V, Karran EH, Boess FG. Quantitative comparison of phosphodiesterase mRNA distribution in human brain and peripheral tissues. Neuropharmacology 2010; 59(6): 367-74.
       [http://dx.doi.org/10.1016/j.neuropharm.2010.05.004] [PMID: 20493887]

[43]   Rabe KF, Tenor H, Dent G, Schudt C, Nakashima M, Magnussen H. Identification of PDE isozymes in human pulmonary artery and effect of selective PDE inhibitors. Am J Physiol 1994; 266(5 Pt 1): L536-43.
       [PMID: 7515580]

[44]   Rabe KF, Magnussen H, Dent G. Theophylline and selective PDE inhibitors as bronchodilators and smooth muscle relaxants. Eur Respir J 1995; 8(4): 637-42.
       [PMID: 7664866]

[45]   Kakkar R, Raju RV, Sharma RK. Calmodulin-dependent cyclic nucleotide phosphodiesterase (PDE1). Cell Mol Life Sci 1999; 55(8-9): 1164-86.
       [http://dx.doi.org/10.1007/s000180050364] [PMID: 10442095]

[46]   Goraya TA, Cooper DM. Ca2+-calmodulin-dependent phosphodiesterase (PDE1): current perspectives. Cell Signal 2005; 17(7): 789-97.
       [http://dx.doi.org/10.1016/j.cellsig.2004.12.017] [PMID: 15763421]

[47]   Chan S, Yan C. PDE1 isozymes, key regulators of pathological vascular remodeling. Curr Opin Pharmacol 2011; 11(6): 720-4.

[http://dx.doi.org/10.1016/j.coph.2011.09.002] [PMID: 21962439]

[48]   Bender AT, Beavo JA. Specific localized expression of cGMP PDEs in Purkinje neurons and macrophages. Neurochem Int 2004; 45(6): 853-7.
[http://dx.doi.org/10.1016/j.neuint.2004.03.015] [PMID: 15312979]

[49]   Sharma RK, Das SB, Lakshmikuttyamma A, Selvakumar P, Shrivastav A. Regulation of calmodulin-stimulated cyclic nucleotide phosphodiesterase (PDE1): review. Int J Mol Med 2006; 18(1): 95-105.
[http://dx.doi.org/10.3892/ijmm.18.1.95] [PMID: 16786160]

[50]   Zhang L, Yang L. Anti-inflammatory effects of vinpocetine in atherosclerosis and ischemic stroke: a review of the literature. Molecules 2014; 20(1): 335-47.
[http://dx.doi.org/10.3390/molecules20010335] [PMID: 25549058]

[51]   Kim N-J, Baek J-H, Lee J, Kim H, Song J-K, Chun K-HA. A PDE1 inhibitor reduces adipogenesis in mice *via* regulation of lipolysis and adipogenic cell signaling. Exp Mol Med 2019; 51(1): 1-15.
[http://dx.doi.org/10.1038/s12276-018-0198-7] [PMID: 30635550]

[52]   Bischoff E. Potency, selectivity, and consequences of nonselectivity of PDE inhibition. Int J Impot Res 2004; 16(S1) (Suppl. 1): S11-4.
[http://dx.doi.org/10.1038/sj.ijir.3901208] [PMID: 15224129]

[53]   Movsesian M, Stehlik J, Vandeput F, Bristow MR. Phosphodiesterase inhibition in heart failure. Heart Fail Rev 2009; 14(4): 255-63.
[http://dx.doi.org/10.1007/s10741-008-9130-x] [PMID: 19096931]

[54]   Perry MJ, Higgs GA. Chemotherapeutic potential of phosphodiesterase inhibitors. Curr Opin Chem Biol 1998; 2(4): 472-81.
[http://dx.doi.org/10.1016/S1367-5931(98)80123-3] [PMID: 9736920]

[55]   Trabanco AA, Buijnsters P, Rombouts FJ. Towards selective phosphodiesterase 2A (PDE2A) inhibitors: a patent review (2010 - present). Expert Opin Ther Pat 2016; 26(8): 933-46.
[http://dx.doi.org/10.1080/13543776.2016.1203902] [PMID: 27321640]

[56]   Huang X-F, Cao Y-J, Zhen J, *et al.* Design, synthesis of novel purin-6-one derivatives as phosphodiesterase 2 (PDE2) inhibitors: The neuroprotective and anxiolytic-like effects. Bioorg Med Chem Lett 2019; 29(3): 481-6.
[http://dx.doi.org/10.1016/j.bmcl.2018.12.018] [PMID: 30554955]

[57]   Degerman E, Belfrage P, Manganiello VC. Structure, localization, and regulation of cGMP-inhibited phosphodiesterase (PDE3). J Biol Chem 1997; 272(11): 6823-6.
[http://dx.doi.org/10.1074/jbc.272.11.6823] [PMID: 9102399]

[58]   Ikeda Y. Antiplatelet therapy using cilostazol, a specific PDE3 inhibitor. Thromb Haemost 1999; 82(2): 435-8.
[http://dx.doi.org/10.1055/s-0037-1615863] [PMID: 10605734]

[59]   Gresele P, Momi S, Falcinelli E. Anti-platelet therapy: phosphodiesterase inhibitors. Br J Clin Pharmacol 2011; 72(4): 634-46.
[http://dx.doi.org/10.1111/j.1365-2125.2011.04034.x] [PMID: 21649691]

[60]   Huang Y, Cheng Y, Wu J, *et al.* Cilostazol as an alternative to aspirin after ischaemic stroke: a randomised, double-blind, pilot study. Lancet Neurol 2008; 7(6): 494-9.
[http://dx.doi.org/10.1016/S1474-4422(08)70094-2] [PMID: 18456558]

[61]   Umebayashi R, Uchida HA, Kakio Y, Subramanian V, Daugherty A, Wada J. Cilostazol Attenuates Angiotensin II-Induced Abdominal Aortic Aneurysms but Not Atherosclerosis in Apolipoprotein E-Deficient Mice. Arterioscler Thromb Vasc Biol 2018; 38(4): 903-12.
[http://dx.doi.org/10.1161/ATVBAHA.117.309707] [PMID: 29437572]

[62]   Bieber M, Schuhmann MK, Volz J, *et al.* Description of a Novel Phosphodiesterase (PDE)-3 Inhibitor Protecting Mice From Ischemic Stroke Independent From Platelet Function. Stroke 2019; 50(2): 478-86.

[http://dx.doi.org/10.1161/STROKEAHA.118.023664] [PMID: 30566040]

[63] Kitamura A, Manso Y, Duncombe J, *et al.* Long-term cilostazol treatment reduces gliovascular damage and memory impairment in a mouse model of chronic cerebral hypoperfusion. Sci Rep 2017; 7(1): 4299.
[http://dx.doi.org/10.1038/s41598-017-04082-0] [PMID: 28655874]

[64] Noma K, Higashi Y. Cilostazol for treatment of cerebral infarction. Expert Opin Pharmacother 2018; 19(15): 1719-26.
[http://dx.doi.org/10.1080/14656566.2018.1515199] [PMID: 30212227]

[65] Houslay M, Schafer P, Zhang K. Phosphodiesterase-4 as a therapeutic target: preclinical and clinical pharmacology. Drug Discov Today 2005; 10: 1503-19.
[http://dx.doi.org/10.1016/S1359-6446(05)03622-6] [PMID: 16257373]

[66] Schaal SM, Garg MS, Ghosh M, *et al.* The therapeutic profile of rolipram, PDE target and mechanism of action as a neuroprotectant following spinal cord injury. PLoS One 2012; 7(9)e43634
[http://dx.doi.org/10.1371/journal.pone.0043634] [PMID: 23028463]

[67] Sasaki T, Kitagawa K, Omura-Matsuoka E, *et al.* The phosphodiesterase inhibitor rolipram promotes survival of newborn hippocampal neurons after ischemia. Stroke 2007; 38(5): 1597-605.
[http://dx.doi.org/10.1161/STROKEAHA.106.476754] [PMID: 17379823]

[68] Atkins CM, Oliva AA Jr, Alonso OF, Pearse DD, Bramlett HM, Dietrich WD. Modulation of the cAMP signaling pathway after traumatic brain injury. Exp Neurol 2007; 208(1): 145-58.
[http://dx.doi.org/10.1016/j.expneurol.2007.08.011] [PMID: 17916353]

[69] Xu B, Wang T, Xiao J, *et al.* FCPR03, a novel phosphodiesterase 4 inhibitor, alleviates cerebral ischemia/reperfusion injury through activation of the AKT/GSK3β/ β-catenin signaling pathway. Biochem Pharmacol 2019; 163: 234-49.
[http://dx.doi.org/10.1016/j.bcp.2019.02.023] [PMID: 30797872]

[70] Tang L, Huang C, Zhong J, *et al.* Discovery of arylbenzylamines as PDE4 inhibitors with potential neuroprotective effect. Eur J Med Chem 2019; 168: 221-31.
[http://dx.doi.org/10.1016/j.ejmech.2019.02.026] [PMID: 30822711]

[71] Chen J, Yu H, Zhong J, *et al.* The phosphodiesterase-4 inhibitor, FCPR16, attenuates ischemia-reperfusion injury in rats subjected to middle cerebral artery occlusion and reperfusion. Brain Res Bull 2018; 137: 98-106.
[http://dx.doi.org/10.1016/j.brainresbull.2017.11.010] [PMID: 29155261]

[72] van Duin RW, Houweling B, Uitterdijk A, Duncker DJ, Merkus D. Pulmonary vasodilation by phosphodiesterase 5-inhibition is enhanced and nitric oxide-independent in early pulmonary hypertension after myocardial infarction. Am J Physiol Heart Circ Physiol 2017.
[PMID: 28986358]

[73] Zayat R, Ahmad U, Stoppe C, *et al.* Sildenafil Reduces the Risk of Thromboembolic Events in HeartMate II Patients with Low-Level Hemolysis and Significantly Improves the Pulmonary Circulation. Int Heart J 2018; 59(6): 1227-36.
[http://dx.doi.org/10.1536/ihj.18-001] [PMID: 30305587]

[74] Lonn E, Shaikholeslami R, Yi Q, *et al.* Effects of ramipril on left ventricular mass and function in cardiovascular patients with controlled blood pressure and with preserved left ventricular ejection fraction: a substudy of the Heart Outcomes Prevention Evaluation (HOPE) Trial. J Am Coll Cardiol 2004; 43(12): 2200-6.
[http://dx.doi.org/10.1016/j.jacc.2003.10.073] [PMID: 15193680]

[75] Arima H,, Chalmers J, Woodward M, *et al.* Lower target blood pressures are safe and effective for the prevention of recurrent stroke: the PROGRESS trial J Hypertens 2006; 24(6): 1201-8.

[76] Wachtell K, Lehto M, Gerdts E, *et al.* Angiotensin II receptor blockade reduces new-onset atrial fibrillation and subsequent stroke compared to atenolol: the Losartan Intervention For End Point

Reduction in Hypertension (LIFE) study. J Am Coll Cardiol 2005; 45(5): 712-9.
[http://dx.doi.org/10.1016/j.jacc.2004.10.068] [PMID: 15734615]

[77]   Hansson L, Lithell H, Skoog I, *et al.* Study on COgnition and Prognosis in the Elderly (SCOPE).
Blood Press 1999; 8(3): 177-83.
[http://dx.doi.org/10.1080/080370599439715] [PMID: 10595696]

[78]   Telejko E. Perindopril arginine: benefits of a new salt of the ACE inhibitor perindopril. Curr Med Res
Opin 2007; 23(5): 953-60.
[http://dx.doi.org/10.1185/030079907X182158] [PMID: 17519062]

[79]   Padma MV. Angiotensin-converting enzyme inhibitors will help in improving stroke outcome if given
immediately after stroke. Ann Indian Acad Neurol 2010; 13(3): 156-9.
[http://dx.doi.org/10.4103/0972-2327.70869] [PMID: 21085521]

[80]   Lewis SJ, Sacks FM, Mitchell JS, *et al.* Effect of pravastatin on cardiovascular events in women after
myocardial infarction: the cholesterol and recurrent events (CARE) trial. J Am Coll Cardiol 1998;
32(1): 140-6.

[81]   Sacks FM, Alaupovic P, Moye LA, *et al.* VLDL, apolipoproteins B, CIII, and E, and risk of recurrent
coronary events in the Cholesterol and Recurrent Events (CARE) trial. Circulation 2000 ; 102(16):
1886-92.

[82]   Schwartz GG, Olsson AG, Ezekowitz MD, *et al.* Effects of atorvastatin on early recurrent ischemic
events in acute coronary syndromes: the MIRACL study: a randomized controlled trial. JAMA 2001;
285(13): 1711-8.
[http://dx.doi.org/10.1001/jama.285.13.1711] [PMID: 11277825]

[83]   Sever PS, Dahlöf B, Poulter NR, *et al.* Prevention of coronary and stroke events with atorvastatin in
hypertensive patients who have average or lower-than-average cholesterol concentrations, in the
Anglo-Scandinavian Cardiac Outcomes Trial--Lipid Lowering Arm (ASCOT-LLA): a multicentre
randomised controlled trial. Lancet 2003; 361(9364): 1149-58.
[http://dx.doi.org/10.1016/S0140-6736(03)12948-0] [PMID: 12686036]

[84]   Ruilope LM, Redón J, Schmieder R. Cardiovascular risk reduction by reversing endothelial
dysfunction: ARBs, ACE inhibitors, or both? Expectations from the ONTARGET Trial Programme.
Vasc Health Risk Manag 2007; 3(1): 1-9.
[PMID: 17583170]

[85]   Böhm M, Schumacher H, Teo KK, *et al.* Achieved blood pressure and cardiovascular outcomes in
high-risk patients: results from ONTARGET and TRANSCEND trials. Lancet 2017; 389(10085):
2226-37.
[http://dx.doi.org/10.1016/S0140-6736(17)30754-7] [PMID: 28390695]

SUBJECT INDEX

A

Ability 10, 25, 34, 36, 83, 86, 104, 138
  highest discriminative 36
  mental 104
Abnormalities 24, 28, 33, 37, 43, 45, 74, 78
  discernible 28
  functional 74
  genetic 43
ACE 129, 131, 138, 140
  and HMG Inhibitors 131
  enzyme 129, 138
  and HMG inhibitors in stroke 140
  inhibitors 138
  inhibitors block 129
Acid 9, 10, 41, 45, 50, 56, 58, 108
  asiatic 58
  ceramidases 10
  docosahexaenoic 108
  gamma-aminobutyric 9, 45
  metabolite octanoic 50
  octanoic (OA) 50
  phenolic 41, 56, 58
Acid amidase 10, 11
  hydrolysing 11
Action tremor 40, 41, 46, 52
  bilateral 41
  generalized 52
Activated 88, 137
  microglial cells 88
  PKA phosphorylates proteins 137
Activation 12, 55, 56, 58, 79, 81, 90, 91, 112, 129, 135, 136, 137
  astrocyte 55
  choline acetyltransferase 58
  glutamate receptor 56
  immune 58
  microglial 81
  pharmacological 136
  platelet 129, 136
Activity 1, 10, 41, 43, 47, 49, 50, 51, 52, 59, 72, 77, 79, 80, 83, 84, 85, 87, 89, 110, 133, 135, 136

antiplatelet 135
antitremor 49
catalytic 133
cerebellar 43
chronotropic 83, 84, 87
inhibitory 136
neuronal 110
neuroprotective 1, 72
penicillin-induced epileptiform 80
protein-aggregate 59
psychomotor 77
rhythmical 47
toxicological 52
Adenylyl cyclase (AC) 131, 134, 137
Adverse drug reactions 140
Adverse effects 1, 13, 14, 49, 51
  cutaneous 13
  unpredictable 1
Agomelatine 78, 80, 82, 83, 84, 85, 86, 87, 88, 89, 90
  analogue 80, 83, 84
  new atypical antidepressant 87
Agranulocytosis 49
Alleviates 81, 108
  epileptiform activity 81
  frequency 81
  tauopathies 108
Alzheimer's disease 55, 59, 103, 104, 105, 106, 108
Amyloid 104, 113
  aggregates 104
  mediated neurotoxicity 113
  precursor protein (APP) 104
Amyotrophic lateral sclerosis 109, 110
Angiotensin 129, 131, 135, 138, 140
  -converting enzyme (ACE) 129, 131, 140
  inactive decapeptide 129
Antibody-dependent cell-mediated cytolysis 7
Anticonvulsant 49, 80, 83
  action 49, 83
  activity 80, 83
Antidepressant effect 76, 81, 82, 84, 85, 86, 87
  plasma melatonin 86

Atta-ur-Rahman & Zareen Amtul (Eds.)

www.ingramcontent.com/pod-product-compliance
Lightning Source LLC
Chambersburg PA
CBHW041708210326
41598CB00007B/580